ROUTLEDGE LIBRARY EDITIONS:
HUMAN GEOGRAPHY

Volume 5

THE GEOGRAPHY OF SEA TRANSPORT

THE GEOGRAPHY OF SEA TRANSPORT

A. D. COUPER

LONDON AND NEW YORK

First published in 1972 by Hutchinson & Co (Publishers) Ltd

This edition first published in 2016
by Routledge
2 Park Square, Milton Park, Abingdon, Oxon OX14 4RN

and by Routledge
711 Third Avenue, New York, NY 10017

Routledge is an imprint of the Taylor & Francis Group, an informa business

© 1972 A. D. Couper

All rights reserved. No part of this book may be reprinted or reproduced or utilised in any form or by any electronic, mechanical, or other means, now known or hereafter invented, including photocopying and recording, or in any information storage or retrieval system, without permission in writing from the publishers.

Trademark notice: Product or corporate names may be trademarks or registered trademarks, and are used only for identification and explanation without intent to infringe.

British Library Cataloguing in Publication Data
A catalogue record for this book is available from the British Library

ISBN: 978-1-138-95340-6 (Set)
ISBN: 978-1-315-65887-2 (Set) (ebk)
ISBN: 978-1-138-95723-7 (Volume 5) (hbk)
ISBN: 978-1-315-66523-8 (Volume 5) (ebk)

Publisher's Note
The publisher has gone to great lengths to ensure the quality of this reprint but points out that some imperfections in the original copies may be apparent.

Disclaimer
The publisher has made every effort to trace copyright holders and would welcome correspondence from those they have been unable to trace.

THE GEOGRAPHY OF SEA TRANSPORT

A. D. Couper

HUTCHINSON UNIVERSITY LIBRARY
LONDON

HUTCHINSON & CO (*Publishers*) LTD
3 Fitzroy Square, London W1

London Melbourne Sydney Auckland
Wellington Johannesburg Cape Town
and agencies throughout the world

First published 1972

The paperback edition of this book is sold subject to the condition that it shall not, by way of trade or otherwise, be lent, re-sold, hired out, or otherwise circulated without the publisher's prior consent, in any form of binding or cover other than that in which it is published and without a similar condition including this condition being imposed on the subsequent purchaser.

The jacket/cover photograph is by courtesy of Overseas Containers Ltd.

© A. D. Couper 1972

This book has been set in Times type, printed in Great Britain on antique wove paper by Anchor Press, and bound by Wm. Brendon, both of Tiptree, Essex

ISBN 0 09 112850 1 (cased)
0 09 112851 x (paper)

TO NORMA

CONTENTS

	Figures	11
	Preface	13
	Glossary	15
	Introduction	19
1	SHIPPING AND CIVILISATION	23
	The northern seas	28
2	MEDIEVAL WORLD-LINKS	32
	Northern expansion	33
	Mediterranean shipping	36
	Widening links, the Indian Ocean and China Sea	38
	Pacific Ocean	41
	New linkages	42
3	BRITISH SHIPPING AND ECONOMIC GROWTH	45
	The domestic coal trade	47
	Development of overseas trade	49
	The advance to steam and iron ships	52
4	WORLD SHIPPING ROUTES	58
	Physical factors	58
	International waterways	62
	Political and legal factors	64

5 TRENDS IN MODERN SHIPPING 69

Demand for shipping	70
Functions of shipping	73
Fleet structure	74
Ownership and growth trends	75
Shipbuilding trends	76
The importance of national shipping	78
Shipping organisation and markets	81
Costs in shipping	84

6 CONVENTIONAL DRY CARGO SHIPPING 90

Tramp shipping	90
The pattern of tramping	92
Tramp earnings	95
Tramp prospects	97
Types of conventional liners	98
Operating cargo liners	100
Conferences	101
Cargo liners: problems and prospects	103
Operating passenger liners	105

7 OIL TANKERS 110

Tanker owners	115
Operating very large crude carriers	116
Oil pollution and tankers	121
Pipelines and tankers	122
Patterns of oil trade	125
Special tankers	128

8 BULK CARRIER AND UNITISED SHIPPING 131

Dry cargo bulk carriers	131
Types of bulk carriers	136
Patterns of trade and freight rates	138
Geographical impact of bulk carriers	143
Unitisation	147
Types of unitised vessels	148
Economics of unit loads	151
Ownership of unitised vessels	154
Impact of unitisation	156

Contents

9	COASTAL AND SHORT SEA SHIPPING	159
	The role of coastal shipping	159
	Types of coastal ships	163
	Ownership of coastal fleets in Europe	165
	The role of short sea shipping	166
	Trends in short sea trade	167
	Trends in short sea shipping	171
	Comparison of the systems	174
	Future developments in short sea trade	176
10	SHIPPING AND THE DEVELOPING COUNTRIES	179
	Freight rates and exports	181
	Freight rates and imports	183
	The drive for national shipping	184
	Development of national fleets	186
	Comments on national fleets	190
	Types of coastal and short sea shipping in developing areas	191
	The role of domestic shipping	193
	Pacific inter-insular trades	195
	Inter-insular shipping in south-east Asia	198
	Index	205

FIGURES

1	World wind systems (January) and ocean routes of European sailing vessels	53
2	Freight rates and laying-up of ships	85
3	Seasonal fluctuations in bulk cargoes (averages 1948–58)	93
4	Average monthly freight rates in grain trade, per long ton (1963–8)	96
5	Main inter-regional movements of oil by sea (1969)	111
6	Oil to Europe by size of ship and route	118
7	Comparative costs of tankers and pipelines	123
8	Disposition of B.P. Group tankers (1 May 1969)	126
9	Seasonal variations in tanker freight rates	127
10	Iron-ore trades (1969)	135
11	Grain trades (1969)	142
12	Ship types, sizes 200–300 m	149
13	Ship types, sizes 100–200 m	173

PREFACE

I started to write this book during 1969 in the Department of Geography, University of Durham. The work was put aside in 1970 when I was appointed head of the Department of Maritime Studies at UWIST. Most of the book has been written at UWIST during 1971.

Where possible, statistics relate to 1971, but all commodity flow diagrams are based on 1969 figures derived mainly from the reports of Fearnley and Egers Chartering Co. Ltd, Oslo. I have also drawn on reports and documents which the Department of Maritime Studies receives from many parts of the world. Particularly useful have been the papers published by UNCTAD, some of the work of the Institute of Shipping Research, Bergen, and the Report of the Committee of Inquiry into Shipping (London 1970) under Lord Rochdale. But much of the material in this book cannot be attributed to particular sources; it has been derived from several years of close contact with shipping and from personal research carried out in developed and developing countries.

I am indebted to my colleagues in the Department of Maritime Studies for reading the manuscript and making corrections. As I gave them little time to do this I cannot in any way share with them my responsibility for any errors which may remain in the book. I am especially grateful to my fellow geographers C. H. Cotter and A. Dredge, to our maritime lawyer W. E. Scobie, naval architect P. W. Penney and economist A. Hassan, while J. J. Evans and B. Thomas checked some of the statistics. I should also like to thank the editor of this series and the staff at the publishers for their understanding and very gentle pressure during my settling-in period at UWIST when so many other commitments had to take precedence over this book.

Finally, I extend my gratitude to my secretary Mrs Pat Buckingham for typing most of the MS, to Mr A. Corner of the Department of Geography, University of Durham, for producing the flow maps and to Mr R. G. C. Cooper of the Department of Civil Engineering, UWIST, for drawing most of the other figures.

Department of Maritime Studies,   A.D.C.
UWIST, Cardiff, December 1971

GLOSSARY

Brig: A two-masted square-rigged vessel usually between 100 and 350 tons.

c.i.f (cost, insurance, freight): This indicates that all charges, freight rates and insurance have been paid by the seller of the goods. The seller usually nominates the ship.

Charter party: This is the document of agreement made between a charterer and a shipowner. It normally contains details of the ship, ports of call, routes, methods of cargo handling, etc. Charters may be 'time', whereby the charterer pays the owner for the hire of the ship at an agreed rate per deadweight ton per month; 'voyage', whereby the ship is hired to carry a certain tonnage between specified ports (under time charters the charterer normally pays for fuel and often for port costs); 'bareboat' (or demise) whereby the ship is taken over for its life or for a very long period and the charterer pays all operating costs.

Clinker-built: Wooden craft with planking in which each strake overlaps the next one below.

Clipper: Square-rigged sailing ship with concave curved bow and rounded stern. Main period 1843–69.

Coir: Rope made from coconut-husk fibres.

Conference: Shipping organisation under which freight rates, schedules and other matters are agreed on between shipowners in respect of particular routes.

f.a.k. (freight all kinds): Where containers or pallets are charged for at a fixed rate irrespective of contents.

f.d. (free discharge): Shipowner meets the expenses of loading, charterer or receiver pays for unloading.

f.i.o. (free in and out): All the costs of loading and discharging are borne by the shipper or charterer.

f.o.b. (free on board): The cargo has to be delivered on board the ship free of all charges. The cost of delivering to the port and loading is thus borne by the seller of the goods. The purchaser usually nominates the ship.

Etesian winds: Northerly winds during the summer in the eastern Mediterranean (Fr. *Vent Eté'sien*).

Lash (lighter aboard ship): A system of transporting loaded barges on board a ship.

Lateen rig: Thought to be of Arab origin. Comprises a large triangular sail attached to a long tapering yard. The peak of the sail extends high above the mast and the yard lies at 45° from the horizontal. Very fast in light winds but can be dangerous under strong gusting conditions.

Loadline zones: These are shown on a chart and are governed by rules which determine the minimum freeboard which a vessel must have within certain sea areas at certain times of the year. Each line corresponding to Summer (S), Winter (W), Winter North Atlantic (WNA), Fresh (F), Tropical (T), Tropical Fresh (TF), is indicated by a horizontal mark on the side of the vessel which must not be submerged in the respective zones.

Standing off-and-on: Sailing vessel approaching the land on one tack and away from it on the other. Used to keep close to a coast without anchoring, or where an anchorage is unavailable.

Stowage factor: This indicates the number of cubic feet which a ton (2,240 lb) will occupy in stowage; increasingly quoted in tonnes (1,000 kilos) per cubic metre.

Tonnages:

Gross Registered Tonnage (g.r.t.) represents the cubic capacity of permanently enclosed space in a ship measured at 100 cubic feet to a ton. Some enclosed space is exempt from tonnage measurement such as double-bottom tanks. g.r.t. is most useful in describing passenger ships and ferries.

Net Registered Tonnage (n.r.t.) is g.r.t. minus the space for crew accommodation, navigation, certain water-ballast tanks and an allowance for propelling machinery. n.r.t. represents true revenue-earning capacity of a ship and is frequently used as a basis for port and canal charges.

Deadweight Tonnage (dwt) is the total load in tons or metric tons (tonnes) of cargo, fuel, stores and ballast which a ship can carry and is usually quoted in relation to the summer loadline. The size of cargo-carrying ships is usually expressed in dwt.

Glossary

Displacement Tonnage (dt) is the quantity of water displaced by a ship and is usually quoted for the full load draught. It may be expressed in tons or tonnes (1 ton = 1·016 tonne).

EXAMPLE

Passenger liner:	length (overall)		147·5 m (484 ft)
	g.r.t.	14,155	(tons)
	n.r.t.	8,137	(tons)
	dwt	2,314	tonnes
	dt	9,774	tonnes
Tanker:	length (overall)		170·7 m (560 ft)
	g.r.t.	14,837	(tons)
	n.r.t.	8,897	(tons)
	dwt	25,233	tonnes
	dt	32,675	tonnes

Trepang (bêche de mer): A species of sea slug (known also as sea cucumber) common in the Pacific. When dried was eaten as a delicacy in China. Thriving trade in nineteenth century and centuries before.

Tumble home: The inward inclination of the sides of a ship above the load waterline (opposite to flare).

INTRODUCTION

The sea has exerted and continues to exert an enormous influence on man. Almost 70% of the globe is covered by sea, more than two-thirds of the world's population live within 300 miles of the sea and most great cities are ports. The world-ocean has a profound influence on climate and hence on population distribution and agricultural activities. The sea has always provided a means of sustenance for many communities, and it is recognised, albeit debatably, as the ultimate reserve of food and chemicals, while the sea bed, particularly the continental shelves, may hold a possible solution to the depletion of mineral resources on land. So far, however, man's most successful use of the sea has been as a highway along which explorations, migrations and vast quantities of goods have flowed. It is the relative ease of spatial interaction coastwise and across the seas which has been, and still is, a principal element in the transformation of places on the world's surface.

This study takes a broad view of the sea as a link between places. The main theme is technological change in sea transport and the impact of this on the character and distribution of economic activities in the world. Technological change in this context means advances in such elements as the design of vessels, building materials, methods of propulsion, navigation and cargo handling. Economic activities relate to commodity production, distribution and exchange.

Maritime transport clearly constitutes part of the distributional sector, but it can equally well be considered as a factor in production. Where, for example, transport (T) is difficult and costly, then labour (L) and capital resources (C) must be drawn from a limited area (i) in the production of commodity (S). If transport is improved, then it may be possible to select labour at L with costs (Li) for part of a process and transport the semi-finished goods to cheaper capital resources at j with costs (Cj) (a hydro-electricity location, for example). It follows that if the costs $Li+Ci>Li+Cj+T$ (i to j) then a transport input in the production of S should be included, all else being equal.

It is self-evident that as transport technology has advanced so the cost of transportation has been reduced. Sea transport costs have steadily diminished over a long time scale and periodically they have fallen sharply. The effect has been to diminish economic distances between places, thus increasing the degree of freedom in the location of production. Technological change has thus a productive function. This may be thought of in any context in terms of a simple model which has a given number of inputs and a quantity of output. Between these there is an assemblage of techniques: if we improve the intervening technology then a greater output may be obtained for the same or fewer inputs. It is, in fact, by raising the power of labour and capital inputs by technological changes in the means of production that most advances in living standards have been achieved.

In the case of sea transport, if shipbuilding methods are improved, or if the capacity, speed and reliability of vessels are increased, then a given transport task can be performed with less capital, labour, or physical risk. So important is this type of technological change that it has been accorded a fundamental place in theories of economic growth. Indeed, the concept of lower transport costs, which broaden the market area and thereby make possible large-scale specialisations based on comparative advantages, goes back to Adam Smith.

By widening markets, or reducing the economic distance between places, many changes are brought about. The effect of improvements in sea transport technology on the character of places is strikingly apparent. One has only to think of the vast areas of relatively low-value foodstuffs and technical crops in North and South America and in Asia, or the agglomeration of raw material using industries at the ports of the industrial nations, to appreciate the extent to which the development of low-cost transport has influenced the location and nature of economic activities throughout the world. Or, to take a specific example, how New Zealand after the coming of the steamship in the 1850s, and the refrigerated vessel in the 1880s, was finally almost completely transformed from its indigenous forest and scrub cover to exotic grasslands to become the low-cost overseas farm of Britain, despite a 13,000-mile separation between producers and consumers.

The impact of technological change is emphasised in this study. It is illustrated by examples which may interest the historical geographer as well as the student of modern international trade.

Introduction

The book appears at a time when more attention is being focused on the sea, and technological advance in the maritime field is very rapid.[1] It may be seen therefore as a small contribution to Maritime Geography, a study which deals with the economic and social aspects of the oceans as well as the physical. Falick offers a useful definition of Maritime Geography in the following terms:

> Maritime Geography is the physical and social study of the sea, its associated rivers, islands, and shorelines, depths and the air above. It is concerned with all that geography normally implies in its emphasis on area, distance, location and interaction of sea phenomena with man.[2]

Much of prehistory, history, and many modern economic relationships may be usefully revealed in the light of maritime studies, and even in this age of the ICBM one finds echoes in power politics and commercial rivalry of the words ascribed to Sir Walter Raleigh:

> Whosoever commands the sea commands the trade; whosoever commands the trade of the world commands the riches of the world, and consequently the world itself.

---

1. Beaver, S. H., 'Ships and Shipping: The Geographical Consequences of Technological Progress', *Geography* 52 (April 1967), pp. 133–56.
2. Falick, A. J., 'Maritime Geography and Oceanography', *The Professional Geographer* 18, 5 (September 1966), p. 284.

I

SHIPPING AND CIVILISATION

The civilisations that evolved in the riverine lands of Mesopotamia, Egypt and north-west India owed as much in their development to trade as they did to agriculture. For the growth and support of cities in the alluvial zones the sustained import of metal, stone and timber was required. These heavy and bulky materials could be delivered in sufficient quantities only by river and sea transport; hence it is in these great drainage basins and adjoining coastlands that we find not only the earliest evidence of urbanisation but also strong indications of a dependence on maritime trade.

There is proof of organised international sea transport in the Persian Gulf during the third millennium B.C. This was clearly centred on Bahrein Island which, lying between Mesopotamia and the entrance to the Persian Gulf, appears as one of the great entrepôts of the period. Vessels from Ur in Mesopotamia carried manufactured goods to Bahrein where they met trading ships from Oman with diorite, copper and timber, and from India with ceramics, gold, silver and ivory. Nothing is known for certain of the actual sea routes in this trade, and the only indication of the level of maritime technology is a seal from Mohenjo-daro on the Indus which shows a vessel with a curved stem and a single mast and sail.[1]

More is known about the ships and trade of Egypt during the third millennium B.C. Egyptian vessels evolved on the Nile; they were built of short sections of local acacia and were flat-bottomed keelless craft with long overhanging stems and sterns. These features suggest development from papyrus rafts and an adaption to loading and unloading end-on to alluvial banks. They carried a large square sail supported by a bi-pod mast stepped forward of amid-

[1] Superior figures refer to end-of-chapter notes.

ships. As a result they could sail effectively only with the wind abaft the beam; but they were ideal for reaching settlements along the banks of the Nile under northerly winds and for returning downstream with the current and powered by oars.[2]

The Egyptian seamen also made coastal voyages to the Lebanon for timber. For this purpose their ships were constructed of longer planks made from imported cedar; and although they retained the form of river craft their sea voyages were made safer by a hawser stretched tight from stem to stern above the deck cargo to substitute for a keel and thus prevent hogging. Outward passages in this trade must frequently have been under oars and close inshore to the Levant to catch the north-going current, while return voyages were made with the northerly Etesian wind of the summer trading season.[3]

Such leading winds were vital for all ancient single-sailed vessels. The Etesian winds which blow seasonally across the pressure gradients of the Mediterranean are drawn also into the Red Sea under the influence of the Asian low-pressure system. There they are channelled between the seaward-facing escarpments of the Red Sea rifts as strong north-north-westerlies. Egyptian seamen evolved a system for using these winds to reach the trading areas of the Indian Ocean. They would leave their home ports in the Red Sea in July and on reaching the Strait of Bab El Mandeb pick up the westerly winds and currents which would carry them towards the main stream of the south-west monsoon. There is no evidence of them having actually crossed the Indian Ocean with their lightly constructed craft; they appear rather to have made for the land of Punt and its entrepôts situated somewhere on the coast of Yemen or Somaliland.

At Punt, Egyptian ships loaded luxury goods, some of which were products of the region but also, possibly, imports from India brought by Arabic and Indian shipping. In October at the onset of the northeast monsoon they would sail from Punt with the south-easterly winds, which are associated with the indrawing of the north-east monsoon by the Sudanese low-pressure system over the middle Red Sea. The same low-pressure belt also draws winds from the Mediterranean, so that difficulty is still often experienced navigating northwards beyond the middle Red Sea convergency zone under sail. As a result the Egyptians built their ports not at the head of the Red Sea or Gulf of Suez but between the latitudes of 20° and 25° North. From the harbours of Bernice and Kosseir goods were transported by overland caravans to ports on the middle Nile where they were transferred to river vessels. This combined use of the pressure systems and land routes provided the main avenues of trade for imports, although by about 2000 B.C. several outward voyages must have been made direct from Thebes by a canal extending from the Nile to the Gulf of Suez across the Wadi Tumilat.[4]

There is no evidence of Egyptian vessels having ventured frequently outside the Levant area of the Mediterranean. It seems that many of the goods which they brought to Egypt from Punt were in fact, along with articles manufactured in Egypt, redistributed in the Mediterranean by more seaworthy merchant ships from the island of Crete. This island was then well situated for such entrepôt functions. It lay on the western margins of civilisation, had good harbours, and had timber for shipbuilding. The Minoan seamen of Crete sought out Mediterranean copper, and other raw materials, for export to Egypt and for home manufactures; and their seagoing ships connected with the land and coastal routes leading to the Mediterranean from Neolithic and Bronze-Age Europe. They built both bulky merchant ships and long fast warships and with these commanded Mediterranean trade until the middle of the second millennium B.C. As late as 1467 B.C. a Minoan vessel is recorded in the timber trade between the Lebanon and Egypt; but by then the brilliant Minoan culture and prosperity, which resulted largely from the predominance of Crete in maritime technology and trade, had given way to that of Greece.[5]

The Mycenean Greeks, who previously had chartered Minoan tonnage, began to spread in the Mediterranean as a maritime people and they established trade links with northern Europe. They pushed the margins of civilisation westwards and thus robbed Crete of its positional advantages. They also adopted the Minoan freighter and improved on it, and they increased the speed of warships by an additional deck of oars. The Greek seamen discovered sailing routes to the Black Sea and initiated an important grain trade. But between 1200 and 1000 B.C. Greek cities were frequently involved in wars. Under these conditions maritime trade far beyond the homeland was often difficult and commercial shipping scarce. As a result much of the Greek commerce fell for a time to the Phoenicians of the Levant coast who were already specialising in trade and in the carriage of goods by land and sea.

The port cities of these latter traders were excellently located on a strip of land between the mountains of the Lebanon and the sea. For centuries the Phoenicians had imported timber and acted as middlemen for commodities arriving by the land routes from Mesopotamia. A painting on Kenamun's tomb at Thebes shows what is probably a Phoenician freighter unloading such a cargo in an Egyptian port; the ship is more heavily built than Egyptian craft, it is keeled, decked, and has high bulwarks suitable for voyages in the open sea.[6]

The Phoenicians sailed widely in the Mediterranean Sea in search of trade, and about 1000 B.C. they entered the Red Sea from a port in the Gulf of Aqaba. Their voyages from this port may have taken them as far as India, and they may even have circumnavigated Africa about 600 B.C., although the evidence for this is slight. The Phoenician seamen certainly reached Sierra Leone by the Strait of Gibraltar and they appear to have been the first of the Mediterranean peoples to trade directly with Atlantic Europe in the search for tin; if any seamen did in fact reach the American continent in ancient times they are likely to have been Phoenicians. So important were they in sea transport that when the Persians conquered Egypt and the Levant in the sixth century B.C. they were left as a semi-independent people to conduct maritime commerce.[7]

Despite the navigational and commercial abilities of the Phoenicians, the Greeks continued to establish seaboard cities at strategic places throughout the Mediterranean. They also built new vessels and by the seventh century B.C. they had shaken off the need to charter Phoenician ships. Seaborne trade had by then become one of the most important economic activities of the Greek world. Following the successful eastern campaigns of Alexander, Greek hegemony was established over much of the Mediterranean, Egypt and the Red Sea. Carthage too flourished at this time and its prosperity was soundly based on sea trade and the command of the western Mediterranean and Strait of Gibraltar.

The Mediterranean seaboard communities received pepper, ivory, sandalwood and muslins by way of the Red Sea, the Canal, and the land routes; and they redistributed these from entrepôts such as Rhodes and Corinth. Their ships carried a proportion of the imported goods direct from East Africa; but shipping in the Indian Ocean must still have remained very largely in the hands of Arabic seamen at this time. For this there appears to be good geographical reasons.

The sailing season in the Indian Ocean was the winter monsoon, from about October to March, with light to moderate north-easterly winds and clear skies. This was the time when lightly built Arab vessels would deliver cargoes to Aden, and other entrepôts. They would then, with lateen sails which allowed them to haul close to the wind, coast back along the Arabian coast before the south-west monsoon strengthened. By contrast Mediterranean seamen would leave ports in the Red Sea in July, would arrive in the Gulf of Aden when the strong summer south-west monsoon was blowing, and return home with the onset of the north-east winds. The rhythm of the seasonal winds and currents thus influenced the pattern of commerce with the Arabs arriving at the ports of trade in early winter when many of the Mediterranean seamen had already set out for home. Some of the Mediterranean traders must have remained at the entrepôts, but it is unlikely that they attempted to compete with the Arabs on sea routes leading to the places of origin of oriental commodities. For them to follow a coastal route to Arabia or India would have meant a long and difficult haul along a low pirate-infested shore, and so the two trading areas must have remained virtually separate entities linked by the great entrepôts in the Gulf of Aden.

Some time between 100 B.C. and the first century A.D. Graeco-Roman seamen did discover how to use the summer monsoon to reach India by a direct route across the Indian Ocean. The *Periplus of the Erythraean Sea* (A.D. 60) attributes this discovery to Hippalus; and it gives navigational directions for the voyages. Vessels would follow the normal sailing courses in the Red Sea and Gulf of Aden, then from the Arabian coast they would run to the Indus with the wind on the quarter. Or, they would make a departure from Socotra Island or C. Gardafui and, keeping the wind abaft the beam, make for Malabar. On the return voyages they would sail from India in December and have the benefit of the winter monsoonal circulation to the middle Red Sea.[8]

The Graeco-Roman ships that made these ocean voyages in all weathers were by the second century A.D. heavily built, broad-beamed vessels of about 250 tons, although in the grain trade from Egypt to Rome vessels may have reached 1,000 tons.[9]

They carried a central mast with a square-rigged mainsail, a triangular topsail, and a sprit at the bow to facilitate steering. They were thus structurally superior to Arab vessels which appear to have remained relatively weak in construction and were frequently laid-up during the stormy south-west monsoon. It seems certain that the discovery of the monsoonal routes not only cheapened goods, by circumventing Arab middlemen, but it brought the East and West much closer and thus accelerated the volume of trade. Strabo records the greatly increased commercial activity of the period with 120 ships per year trading to India with cargoes which included Mediterranean wine, glass and metals.

Thus by the second century A.D. international sea transport linked the Mediterranean with Atlantic Europe to the west and India to the east, while at both ends of these links maritime and land connections were made with remote areas. This commerce was built up over several millennia by the discovery and use of the wind systems and by advances made in maritime technology which allowed new routes to be followed. The design of Mediterranean ships which emerged at this time was to persist for centuries; and many of the characteristics of international sea transport which came into use between 500 B.C. and A.D. 200, such as shipowners' associations, sharing of voyage risks, charters, and many clauses and marine insurance practices, were to pass into modern commerce.[10]

Little is apparent of the patterns of sea trade beyond India during the first century A.D. Moslems are known to have settled on the coast of China before A.D. 50, where they acted as middlemen between foreign seamen and the Chinese, but the information is fragmentary. There are even fewer documents relating to maritime commerce at the other end of this link beyond the western Mediterranean. The explorations of Pytheas to the northern seas in 300 B.C. give little indication of trade. However, there is sufficient archaeological evidence to sketch in some aspects of the maritime relations of northern Europe during prehistoric times.

The northern seas

There is no doubt that movement by sea in colonising, fishing and trading must have taken place very early in the prehistory of northern Europe, but there is only limited evidence of this before the Neolithic colonisation in the third millennium B.C. Wooden paddles from Germany and a dugout from Scotland are all that remain of what could have been quite extensive Mesolithic coastal navigation.

Of the Neolithic people who entered Britain about 2500 B.C. it is the so-called Megalithic group who show the strongest attachments to the sea. The distribution of their collective family tombs from Sicily to Scandinavia marks the settlements and trade routes in the maritime movement of these culturally related communities. The tombs are, very significantly, found clustered on peninsulas and islands, and frequently they command portages across headlands where weather and tidal conditions made sea passages difficult.[11]

The maritime values of the Megalithic farmers and fishers are particularly emphasised by the remains of their widespread trading activities. All the peoples of the New Stone Age required durable stone tools for forest clearing and they obtained these from quarries and factories most of which have been discovered close to the sea. It is clear from the existing distribution of stone implements and from petrological evidence that the trading vessels of Neolithic times carried shipments of roughed-out heavy igneous stone axes throughout the Atlantic seaboard; from north-west England to Scotland and southern England; from Wales to as far as Jersey; and from northern Ireland to Scotland and England. Similarly, in Scandinavia flints from the Cretaceous areas of Jutland and southern Sweden were transported over 900 miles through the Baltic to rivers entering the Gulf of Bothnia, and some of these flints were then carried overland to Norway and once more redistributed coastwise.[12]

It is difficult to establish the type of craft used in this trade during Neolithic times. The initial colonisation of western Britain and the 300-mile voyage made by colonisers from Scotland to Norway must have required vessels sufficient to transport viable family units, food, water, stone implements, seed and livestock. The weather under the dry conditions and the probable easterly breezes of the sub-Boreal period would have been favourable to voyaging, but it does seem unlikely that the small wooden dugouts regularly found in Neolithic sites would have been their only means of transport.

Neolithic vessels were probably large skin and wicker curraghs, such as those the Mediterranean seamen saw sailing between Brittany and Ireland in 600 B.C., and which remained in use in Ireland, as ocean-going vessels, until the seventeenth century. But there may also have been larger wooden ships in Neolithic Europe, for, as Lindsay Scott says, 'it would be surprising if the settlements which introduced the chambered tomb failed to transmit Mediterranean shipbuilding techniques up to that standard'.[13]

Be that as it may, it is certain that the movement of people by sea from the Mediterranean, and their contacts through trade with former homelands, gave rise to new commercial activities in Europe and eventually to the diffusion of bronze metallurgical techniques into the Stone-Age cultures. With the actual settling of Bronze-Age people in Britain, about 1900 B.C., there was a further expansion of trade and a concomitant spread of more advanced technology. Bronze-Age merchants traded in Cornish tin, Irish gold and copper and British and European amber. The long period of mutual exchange of goods between Britain and the Mediterranean civilisation is indicated by British amber and gold articles found in Minoan Crete and Mycenean Greece, and the hoards of Egyptian, Minoan and Mycenean goods which have been unearthed in southern England. The element of sea transport which is implicit in some of these exchanges is further testified to by a Bronze-Age hoard containing goods with Hiberno-British, Biscayan-French, and Mediterranean affinities recovered from a harbour in Portugal.[14]

The ships of these Bronze-Age traders appear to have been of three main types. The oak dugout of the Neolithic period to which a sheer strake had been attached, the Neolithic skin curragh, and a new long-planked type of vessel, as depicted in the petroglyphs of Tanum in Sweden and in the Als ship excavated in Denmark. Very broad-beamed craft are also indicated by gold votive models of ships from Denmark, and the Bronze-Age peoples of Britain must have had considerable knowledge of maritime technology since they appear to have possessed vessels capable of transporting the enormous stones of Stonehenge by sea from Pembrokeshire to the River Avon.

Many of the Iron-Age peoples who followed the Bronze-Age settlers to Britain from 800 B.C. onwards came also by a long sea route. There was a marked climatic deterioration at this time, but these Celtic peoples built substantial ships capable of voyaging in boisterous northern seas. The vessels of the Veneti tribe of southwest England and north-west France, for example, were described by Caesar in 56 B.C. as having flat bottoms for grounding in the tidal coastal waters, high bows and sterns fitted for use in storms, hulls made entirely of oak, beams a foot wide fastened with iron bolts, anchors secured by chains, and sails of hide or leather.[15]

The Veneti were seafarers and traders who commanded the coastal and cross-channel commerce of southern Britain for several centuries before the Roman invasion; they were also skilled shipbuilders and their distinctive ships equalled if not surpassed those of Mediterranean construction. Some of the ports of the Veneti were undoubtedly located on the south coast of Britain, but the maritime complex of villages near Glastonbury on the Severn is also thought to be associated with these sea traders. The descendants of the Veneti may have continued to trade in northern seas during Roman times; certainly the great expansion in agriculture, mining and industry under the Romans stimulated more seaborne commerce and carried elements of Mediterranean civilisation further northwards into Scandinavia.

Man's ability to use the rivers and seas, not only as a source of sustenance but as a means of long-distance transport to obtain scarce commodities, played a vital part in the extension of the geographical division of labour, the development of areal specialisations and the rise of port urban centres as places of exchange. The first application of non-human motive power in the form of a sail, and the growing dependence of spatially separated communities on the products of other areas, stimulated further technical and scientific skills in shipbuilding and navigation, and led to a greater diffusion of people, material-culture and ideas along the sea routes. These movements were especially strong from the Mediterranean to north-west Europe, but they extended as far as China and south-east Asia, suggesting that very ancient trade had many of the characteristics of a world trade.

The evidence for distant links is still, of course, fragmentary, but there does appear to be a gradual unfolding through archaeology of a greater geographical area of contact by sea than was hitherto thought probable for prehistoric times. These relationships between the sea as a routeway, the development of shipping, and the cultural and material advance of civilisation occupy a fundamental place in the historical geography of the ancient world.

1. Oppenheim, A. L., 'The Seafaring Merchants of Ur', *Journal of the American Oriental Society* **74** (1954).
2. Hornell, J., *Water Transport, Origins and Early Evolution* (Cambridge 1946).
3. Pritchard, J. B., 'A trip to the Lebanon for Cedar', *Ancient Near Eastern Texts* (Princeton 1950).
4. Breasted, J. H., *A History of the Ancient Egyptians* (London 1908).
5. Glotz, G., *Aegean Civilization* (New York 1925).
6. Faulkner, R. O., 'A Syrian Trading Venture to Egypt', *Journal of Egyptian Archaeology* **33** (1947).
7. Horden, D., *The Phoenicians* (London 1962); Hadi, Hasan, *Persian Navigation* (London 1928).
8. Schoff, T. R. (Ed.), *The Periplus of the Erythraean Sea* (London 1912).
9. Casson, L., *The Ancient Mariners* (London 1959).
10. Toutain, J., *The Economic Life of the Ancient World* (London 1912).
11. Kirk, W., 'The Primary Agricultural Colonisation of Scotland', *Scottish Geographical Magazine* **73** (1957).
12. Clark, J. G. D., *Prehistoric Europe* (London 1965).
13. Scott, L., 'The Colonisation of Scotland', *Proceedings of the Prehistoric Society* **17** (1951).
14. Clark, ibid.
15. Handford, S. A. (trans.), Caesar, *The conquest of Gaul* (London 1958).

2

MEDIEVAL WORLD-LINKS

The disintegration of the Roman Empire in the west brought to an end a period of unified control over many of Europe's main channels of trade. Land routeways from the Mediterranean were thereafter blocked frequently by incursions of barbarian tribes. But sea transport endured, and the Byzantine Empire survived to police the eastern Mediterranean. It was the continued existence of shipping, Pirenne maintained, which kept the economic organisation of the Mediterranean world alive in these critical times.[1] Maritime commerce also survived the decline of Roman power in northern Europe, for Kent retained some commercial relations with Denmark, Frisia, the Rhine and the Celtic West.[2] But these were the Dark Ages and documentary details of trade are scanty.

In the Mediterranean Byzantine shipping continued to serve a relatively wide region. But in northern seas the ships and commerce can be identified, from archaeological evidence, only within certain restricted trading areas. The Nydam boat of the fourth century, for example, while it was an advance on earlier Scandinavian models, in that it possessed oars instead of paddles and had planks joined by iron nails instead of stitching, was still without sails, built very low amidships, and was probably confined to voyages in the short seas of the Baltic. The Utrecht ship also shows local characteristics, with a broad flat keel very suitable for trading along the indented and shallow sandy coast of the Netherlands, but less suited for sea voyaging.[3]

A possible revival of more long-distance trade may be associated with the Anglo-Saxon Sutton Hoo vessel of the seventh century. This appears to have been a seagoing ship with a higher bow, broader beam, and more stability than the craft of the early post-

Roman era. The goods buried with this ship may also testify to trading connections between the continent and Anglo-Saxon England, if not further.[4] In the west of Britain the deep-sea curragh traders appear likewise to have been making ocean voyages. These Celtic seamen certainly preceded the Norse to Iceland, possibly by several centuries; and the sailing Scaphas of Gallic traders continued to ply to the coast of Spain as in Roman times.[5] Sea transport thus remained active during the Dark Ages, and later; and shipping must have continued to provide at least a tenuous series of links between the Mediterranean and northern Europe throughout this period.

It was not in fact until the invasion of the Mediterranean mainland and islands by the Arabs in the seventh century that the traditional trading unity between north and south was for a time effectively severed. The impact and extent of the Arab invasion were remarkable, notably in the Levant, the ports of which had previously been controlled by the Byzantine Empire, and thus this important supply line for oriental products to Europe was cut. The Arabs then moved across the Mediterranean, reducing to a trickle the flow of trade to ports such as Marseilles, and finally interposed a unified, but hostile, power between northern and southern Europe and the Middle East.

One result of the Islamic invasion was to bring into more significance the trade routes by land and river from Byzantium eastwards to the Caspian Sea and north-westwards to reach the river valleys leading to North Sea ports. However, by the seventh century this northern coastline was under attack from sea forces in the form of Norsemen; and by the eighth century Norse raiders had contributed to a decline of much of the precarious commerce of the southern North Sea entrepôts. The medieval commercial western world thus shrank once more, under the impact of the Arabs from the south and the Vikings from the north, to 'an economy of no markets' with little direct trade between the regions.[6]

Northern expansion

By the tenth century the Viking seafarers had turned more markedly from raiding to trade and colonisation. Indeed, Brønsted considers that it was the quest for trade which was, possibly, the most crucial factor in the ultimate Viking expansion.[7] They pioneered the Northern Way, that great trade route along the coast of Norway to the White Sea. They also followed all-water routes from their lake ports in central Sweden, across the Baltic and along the principal rivers into Russia and central Europe to reach the entrepôts of oriental trade in the Black Sea, and beyond. During the eighth and ninth centuries the Vikings settled in the coastal and island areas of Britain and colonised Ireland and northern France. In A.D. 870 they opened the maritime route to Iceland, which they probably learned of from the Celts; and by the end of the tenth century they were trading to Greenland and ultimately reached North America. Their voyages took them also to Spain and into the Mediterranean Sea in search of oriental products.

The forging and maintenance of this vast Viking commercial empire was made possible by the superior shipping of the Norsemen during the period from the seventh to the twelfth century. Details of the construction, speed and versatility of their longships have been well documented, but by the ninth century they had also evolved the knarr, or hafskip, for the purpose of ocean trading. The longships were unsuited to this latter function; they were shallow-draughted, narrow ships of war depending on great numbers of oarsmen for military power and speed. The ocean trading vessels were by contrast shorter, broader-beamed and deeper craft, powered by a large square sail; consequently they required only a small crew and had greater cargo capacity.

The trade of the Vikings was multifarious; cargoes from the White Sea, Iceland and Greenland comprised furs, skins, hides, horn, tusks, oil, dried fish, woollen cloth, butter and falcons. These they sold throughout medieval Europe; and they traded to Scotland for meal, salt, and iron; to England for grain, malt and cloth; to Flanders for cloth, wine, jewellery and spices; and to Ireland for flax. The western seaboard, and particularly the port of Dublin, was important for the Norse trade. From this area they transhipped luxury commodities brought from Spain and the Mediterranean.

The Norse cargo ships at the height of this trade were still undecked vessels. Cargo was stowed amidships, covered with ox hides, and on top of this a lifeboat was lashed. Passengers were also carried and merchants generally travelled with their goods—often working the ships as crew in lieu of freight rates. The master, as was the medieval custom, was frequently a part-owner of the ship and the cargo.

Such vessels were sturdy but their voyages were largely determined by the seasons of the northern seas. Sailing was done primarily in the summer months between April and September. Very few ships would venture on the longer voyages later than September—when the skies became overcast, days shorter and storms more frequent. Normally, they would arrive in Iceland or Greenland in early summer or autumn and would remain there trading throughout the winter, returning to Europe in the spring.[8]

The maritime commerce of the Vikings made a very significant contribution to the economic development of Europe in the late Middle Ages. Their activities helped revive and regularise the whole trading milieu of the period, and they introduced uniform commercial laws which were observed in many parts of the northern world.[9] But Lewis cautions against overestimating the importance of the Vikings in this respect, for there were many other merchant vessels active on the northern seas by the twelfth century. They not only competed with the Vikings commercially but precipitated the decline of their political power.[10]

The revival of commerce to which the Vikings contributed so much in the tenth century had in fact greatly stimulated Flemish, German and Celtic shipping. By the eleventh century there was already a marked reduction of the Viking trading empire with the loss of Ireland. The Norse merchants were further weakened by climatic deterioration in the thirteenth and fourteenth centuries, which disrupted their trade to Iceland and Greenland and brought poor harvests to Norway; and again by the Black Death which swept through Scandinavia during the fourteenth century. All these circumstances combined to produce a depressing effect on the Norwegian economy and contributed to political upheaval. This brought with it indebtedness to the German merchants of the Hanseatic League—a powerful group commanding the financial resources of German sea and river ports, which acquired control of the Norse trade from Bergen.

In Flanders also there had evolved substantial trading cities, such as Ypres, Ghent, and Bruges. These developed in low-lying somewhat infertile areas, being excellently situated within a network of navigable waterways. In addition, local wool and weavers made the towns attractive to traders and they established regular overland communications with Venice and Byzantium, as well as sea connections with the rest of Europe. In England the Cinque Ports formed an association similar to the Hanse, and like the merchants of Flanders and the Hanse they owned ships and sailed these in convoy on trading voyages to northern Europe.

By the thirteenth century the full flood of commercial revival was also under way in southern Europe. Genoese merchants had resumed trading along the western seaway route from the Mediterranean to the ports of Flanders and southern England. The luxury goods such as silk and ivory which now flowed freely from the Orient depressed the value of tusks, hides and other northern products, and this likewise reduced commercial incentives in the trade to Iceland and Greenland. As a result Scandinavia, which had done much to stimulate trade, was gradually becoming an economic backwater.

In addition to the changing trading and political conditions of the late Middle Ages there was one other major reason for the decline in

Viking commerce: this was the extreme conservatism which had set in on matters of maritime technology. By the thirteenth century medieval economic life had quickened, cities had grown in size and there was a greater demand for bulky products, such as grain and timber, as well as for luxuries and for passenger travelling facilities. Both western European and Mediterranean merchant shipping responded to these demands by increasing the carrying capacity of ships and improving their reliability and sailing qualities, but little change took place in the designs of Scandinavian vessels.

Viking merchant ships, like most northern cargo vessels, were in the twelfth century double-ended undecked craft of 50 to 80 tons, fitted with a single mast and a square sail. By the thirteenth century the vessels of Germany and Flanders had increased in size to around 100 tons and were broad-beamed with high forecastles and aftercastles. Their rigging had also improved to allow sailing closer to the wind, and the high bow gave them the added protection required for this practice. By the beginning of the fifteenth century there had been an acceleration of technological change; the side steering oar had given way to the stern rudder, which greatly facilitated manœuvring, and many ships were by then fitted with three masts, two square-rigged and a lateen on the mizen. They could sail on the wind, and some, such as the 600-ton Flemish carracks, had adopted the Mediterranean carvel type of hull which gave added strength for the carriage of heavy guns on deck.[11]

The Scandinavian ships were thus completely outclassed in northern seas by the vessels of the Germans, Flemish and Dutch. The English, at this time, appear to have been curiously ambivalent in relation to shipping. They depended greatly on maritime trade but their own ships were relatively small and few in number, a situation which they remedied only in the late sixteenth century.

Mediterranean shipping

When the centre of power of the Roman Empire shifted to Constantinople in A.D. 330 Roman control was maintained over land routes to Persia and the seaways to the Aegean and Levant. Constantinople thus became the great entrepôt of the Mediterranean. Later invasions from the north forced Italians eastwards into the marshes of Venice where they also developed a sea-trading economy initially serving the needs of Byzantium.

Byzantium, although commercially powerful, was constantly in conflict with Persia over trade and the control of Syria and Egypt. It was probably the weakening of the two major powers in this way which allowed the Arabs to spill out from their relatively heavily populated desert environments and successfully overcome both forces. In the early seventh century they occupied the Levant, in A.D. 660 they conquered Sicily, and by 711 had invaded Spain and were in possession of the whole of the southern Mediterranean coast.

Apart from the seamen of the south Arabian and Persian Gulf coasts the Arabs were basically a land people. Nevertheless, by using Syrian and Egyptian sea forces and improving their own vessels they defeated the Byzantine navy in 655 and almost closed the western Mediterranean Sea to European commerce. They reopened the canal between the Nile and the Red Sea and reoriented the grain trade of Egypt in an eastwards direction. But all western trade did not cease permanently, for through the intermediary services of Jewish merchants and seamen some of the silks and spices arriving in Egypt from the Orient did find their way to Venice and Constantinople.

By the mid-eighth century Byzantium re-established considerable control over the north-eastern part of the Mediterranean Basin, and continued to trade along Black Sea and land routes for the products of Asia. The rising Venetians in turn concentrated some of these cargoes of oriental origin at their port, and they conducted a trans-Alpine trade with northern Europe. In the late ninth century armed Venetian galleys were defying Saracen pirates and were sailing in convoy to trading places in the Aegean and eastern Mediterranean. But the resumption of full-scale trade in the Mediterranean had to await the concentrated attacks on the Arabs by the Christian powers during the eleventh century. By then Byzantium was under pressure from the Turks and her overland trade routes were frequently closed. This gave further opportunities to Venice and the other maritime cities of Italy, and gave a strong incentive to reopen by force the ports of the Levant as centres of commerce between east and west. The Italian cities of Venice, Genoa and Pisa had a vested commercial interest in the success of the armed Crusades. Their vessels carried the Crusaders and their supplies to the Levant; and under the protection of these forces they opened once again the Orient trade.[12]

In time the Moslem Empire disintegrated into a number of independent units and its power weakened. By the thirteenth century European maritime supremacy had returned to the Mediterranean in the form of highly armed Venetian galleys which sailed regularly to Egypt and North Africa. These were warships; but the Venetians developed also a merchant galley for the eastern Mediterranean trade. Here, in fact, were the first true cargo liners. They were built by the state and chartered to merchants who ran them to a fixed schedule. The ships were fast and were flexible in manœuvring, they carried 100 to 150 oarsmen and were fitted with three masts, all with lateen sails. Such galleys proved expensive to construct and operate and they had little cargo space, but they specialised in the high-value low-bulk products of the Orient for sale in Venice.

The pattern of trade of the Venetian galleys conformed far less to the wind systems than it did to the demands of the market. The ships would sail in convoy from Venice to Alexandria in July and August, and return via Rhodes and Crete in November and December—carrying cargoes of silks, damask, dates and spices in time for onward distribution to the Christmas fairs of Italy and northern Europe.

There was less regularity, or cargo specialisation, in the trade of the Genoese. Their ships included the carrack type already referred to in connection with Flanders. They were high-capacity broad-beamed vessels that normally voyaged under sail only, carrying the heavier and bulkier lower valued goods such as cotton, dyes, timber, and grain, of the Mediterranean trade; consequently they had less need for overmanning in the interests of defence and speed.

In the thriving trade which eventually developed to Flanders and the Cinque Ports the Venetians, Genoese and Florentines combined in their ship construction some of the characteristics of both the fast galley and the bulkier carrack. The resulting Atlantic merchant galleys were large, stable, broad-beamed vessels of 500 to 1,000 tons. The number of oarsmen was reduced on these ships and additional sails were used. They were extremely seaworthy craft, for, as Mallett points out, they operated to England and the continent during the depth of winter and suffered few losses.[13] Winter sailing must have increased enormously the productivity of Italian shipping; and the factors of added reliability and safety must have encouraged a further expansion of commerce and the increased movement of merchants during the late Middle Ages. The wealth accumulated from these maritime enterprises supported the great cities, arts and culture of Italy: until, that is, their monopoly of the eastern trade was challenged by the opening of a new oceanic trade route by the Portuguese around the Cape of Good Hope during the early sixteenth century.

Widening links, the Indian Ocean and China Sea
The luxury commodities distributed in the Mediterranean, northern Europe and the north Atlantic during medieval times had their origins in a vast oceanic-linked area encompassing Africa, India, south-east Asia and China. The economic centre of this was China, the most advanced and civilised country in the world after the decline of Graeco-Roman power. However, this region was knit together commercially for many centuries not by the Chinese themselves but by Arab seamen from the south Arabian coast.

Like the Vikings the Arabs occupied a difficult physical environment in proximity to a richer world. They too engaged in pirate activities, raiding, colonising and trading. But unlike the Vikings they lacked good timber for shipbuilding. Their vessels were sparingly constructed of imported teak—or occasionally built at sources of good timber in India or the Maldive Islands. Planks were stitched together with coir and the seams covered with pitch. They were not strong ships and on long-distance voyages tended to avoid sailing during the seasonal periods of heavy seas.[14]

What the Arabs lacked in the way of ship strength they made up for by their knowledge of winds, currents and astronomy, and by the development of the lateen sail which gave speed and manœuvrability to their ships under adverse winds. The lateen, which converts lateral wind pressure into a forward force, allowed them to make long reaches across the Indian Ocean holding their ships within four or five points of the wind.

In the trade to the Far East Arab ships would leave home ports towards the end of November and under the north-east monsoonal winds would reach Malabar in about a month where they would commence trading. After a few weeks they would sail for north-west Malaya, often via Ceylon, and again engage in trade; then, with the last of the north-east wind, some time about March, would pass through the Malacca Strait and on to Palembang or other ports of trade in Sumatra. When the south-east monsoon came away in April they would finally make for Canton.

The time of the markets at the various ports in this Far Eastern trade was undoubtedly regulated by the monsoon. For example, when ships arrived at Canton about May their cargoes of linen, cotton, wool, rugs, shell, horn, ivory, metal, sandalwood and spices from Europe and Africa (and collected *en route*) were held by the authorities until the last of the season's fleet had berthed. Then trading would commence and would end at the onset of the north-east monsoon in November. The fleets would then sail southwards, and after beating through the Strait of Malacca would, all being well, have the benefit of a north-east wind until they reached the coasts of Arabia or the Red Sea at the start of the south-west monsoon. Their homeward cargoes included silk, camphor, pepper and sugar from China and spices from Indonesia.[15]

At the ports of this trade in China and south-east Asia there were permanent colonies of foreign merchants and seamen occupying their own quarters of the towns, much as in the staple ports of Europe. In this activity the Arab merchants were also dominant; in fact during the seventh century they were reported as being the only foreigners in Canton. But the entrepôts of south-east Asia were served also by Indian, Burmese and Persian merchants and shipping. There were in addition to these deep-sea traders local south-east Asian seamen and craft, including the fast proas of the Indonesians, engaged in sailing the archipelagos collecting small quantities of spices and sandalwood from a myriad of islands for concentration at the principal entrepôts,[16] and possibly connecting with northern Australia and the western Pacific islands. This feeder arrangement was repeated on the coast of eastern Africa where local vessels would bring gold, ivory, and slaves from the southern part of the continent to the Arab marts;[17] and again in north China with the local trade from Japan and Korea.

The Chinese themselves came relatively late into ocean trading, although their river and coastal traffic was very ancient, and there are references in Chinese literature of visits to Japan during the Han dynasty (207 B.C.–A.D. 220). They explored the seas much more widely in the seventh and eighth centuries A.D. and established ambassadors in many countries. By the twelfth century they were remarkably active traders and their junks must have been amongst the largest class of vessels afloat. Many medieval writers, Chinese, Arab and European, refer to these ships. Marco Polo describes them as decked, double planked, double bilged against stranding and fitted with watertight bulkheads; they had, he says, sixty cabins for the merchants, stowage space for 5,000 to 6,000 buckets of pepper, and a crew of 150 to 200.[18] Chinese ships of the period had three or four masts, large batten sails, and fitted rudders; but their hulls were variously designed according to the physical conditions of the coasts and ports to which they belonged.[19] They traded to south-east Asian, African and other ports of the Indian Ocean from the thirteenth to the fifteenth century. It was also, by then, the practice for merchants arriving in south-east Asia to tranship their goods, and themselves, from some of the smaller Arab vessels to Chinese junks for the voyage to Canton; for such large ships, in addition to being more seaworthy and towing a reserve vessel, were well equipped for encounters with Tonkinese and Japanese pirates.

The long peninsulas and archipelagos of south-east Asia thus represented in medieval times a physical, economic and cultural interface between the Indian Ocean and western Pacific. Here Arab, Chinese, Indian and other merchants gathered at major ports awaiting the seasonal shifts in the winds and the arrivals and departures of local shipping. A situation not unlike that at other entrepôt areas in the western Baltic, Flanders, the Aegean, or the southern end of the Red Sea, to which this oriental trade was connected by a series of shipping links.

But before eastern goods could reach the distant western entrepôts

Medieval world-links

through these channels of trade numerous costs were incurred. Fires in the warehouses of Canton often destroyed stock, ships were lost to pirates, reefs, or typhoons; then there were gifts to be presented to rulers *en route* and taxes to be paid. In Canton alone three-tenths of a cargo of a foreign ship had to be surrendered as taxes.[20] Finally, there were the profits of the various shipowners and merchants, the transit taxes imposed by Egypt, the profits extracted by the Venetians, and the profits of the north European middlemen. All such transfer costs and charges raised the price of oriental commodities in Europe and gave added impetus to the quest for a more direct route to the Spice Islands during the fifteenth century.

Pacific Ocean

Once the archipelagos of south-east Asia were cleared ships came to the edge of what the Chinese called the Great Eastern Ocean. Eastwards was the vast expanse of the Pacific from which the north-east and south-east trades blow towards south-east Asia. Such winds would appear to give little assistance to further eastward sailing. But low pressure over the Australian continental area during the southern summer appears to draw the north-east trades across the equator as the north-west monsoon, and in winter winds associated with the anticyclonic system over Australia blow towards the Asian low as the south-east monsoon. These were the winds used by Chinese ships in their trade to Indonesia, and utilised in turn by Indonesian vessels on voyages to northern Australia for trepang (*bêche-de-mer*), a delicacy which they traded to the Chinese. This ancient trade was conducted by large Indonesian proas and was still going on when the British occupied Australia in the eighteenth century.[21]

There appears also to have been some two-way contact between south-east Asia and some of the islands of Micronesia—in spite of the seemingly adverse wind conditions for eastwards sailing. In actual fact, the westerly gales which periodically occur in this area about January, the Micronesian's knowledge of the Pacific countercurrents, and their use of fast lateen-rigged ships, would have ensured the Pacific island traders of return passages from south-east Asia. Asian pottery discovered in the Marianas dating from pre-European times and glass beads found on Palau strongly suggest such trading contacts with the Philippines, and indirectly with China and Persia. Trepang from the Pacific coral reefs was again the most likely commodity exchanged for these highly prized ornaments.[22]

Apart from such tenuous links near the rim of the ocean the south Pacific world, including Australia, New Zealand and the great scatter of islands, must have remained isolated from the mainstream of maritime commerce which built up in the Middle Ages. It was clearly too remote from Asia, and Asia too rich in resources, to be interested in extensive Pacific ventures.

The Pacific islanders did, of course, have their own systems of maritime trade within and between archipelagos. They made some impressive voyages, and their ships could rival European craft of the eighteenth century in size, and excel them in speed. This oceanic traffic distributed and redistributed island products and communities throughout the Pacific Ocean before and during the Middle Ages.[23] The islanders acquired navigational skills and detailed geographical knowledge and as a result specialised products could be exchanged between distant places.[24] The size and speed of Pacific ships, as recorded by the early European explorers, bear adequate testimony to the high level of maritime technology that could be attained by a sea-oriented people even at a Neolithic level of material culture.

New linkages

The ancient channels of trade from the Orient via the Mediterranean to Atlantic Europe still served their purpose at the beginning of the fifteenth century. However, the capture of Constantinople by the Turks in 1453 destroyed one of the major links between East and West. The Ottomans also succeeded in blocking the routes across Syria; leaving the Isthmus of Suez as the only means of access for oriental goods. This transit area became subject to even higher dues imposed by Egypt. These events brought economic difficulties to the Mediterranean trading cities, and caused the western European powers to intensify their search for a new sea-route to the sources of eastern trade.

The westwards voyage of Columbus in 1492 did not provide the alternative route; but in 1498 Vasco da Gama reached India by way of the Cape of Good Hope. The Cape route not only freed the merchants of Atlantic Europe from charges for the Suez Isthmus transits, and those paid to Mediterranean intermediaries, but also enabled them to bypass the ancient link provided by the Arab traders in the Indian Ocean. The Portuguese defeated Arab naval resistance to their encroachment and established their own ports of trade in India, south-east Asia and at Macao in China.

Spain was equally intent on reaching directly the riches of the Orient. By the Treaty of Tordesillas the unexplored and conceptually flat world was divided between Portugal and Spain. Lands to the east of the Tordesillas line, which extended along the meridian of approximately 37°W, went to Portugal and those to the west fell to Spain. The Spaniards sought, therefore, an approach to the Indies by sailing westwards beyond America. In 1521 the Spanish expedition initially led by Magellan rounded South America, sailed across the Pacific to the Spice Islands and returned to Spain in 1522.

Magellan's discovery of the Pacific route led to the establishment of new entrepôts in central America. From there the Spaniards traded Mexican silver to Manila where they collected the products of south-east Asia and China and returned with these to Mexico. Eastern goods unloaded in Mexico were, by the mid-sixteenth century, carried by pack trains across the Isthmus of Panama, and in the early seventeenth century by river and canal across the Isthmus San Pablo, and forwarded to Europe. By then Spanish seamen had discovered methods of sailing to and from Manila by utilising the winds of the Pacific high-pressure system; methods which mirrored the trading patterns then in use in the Europe–West Indies trade around the Azores high. The galleons would sail outwards from Acapulco into the north-east trade-wind zone and across the vast expanse of the Pacific to Manila. For the return voyage they would press north-eastwards from Manila to reach the prevailing westerlies which would carry them to northern California, and from there they sailed southwards under northerly winds to Mexico.

The fifteenth century thus saw the final linkages in the commercial ties between Europe and the Orient. The new sea-routes which by-passed the ancient trading links were made possible by advances in the size and rigging of western ships, especially the Portuguese caravel which was capable of sailing close to the wind, the revival of Ptolemaic geography, and the use of the compass and new astronomical navigation.[25] One of the significant aspects of these developments was that for the first time since the Graeco-Roman period an area which contained the principal markets for oriental goods was now in direct contact with the sources of supply. This ability to control the ports and channels of trade directly by vessels from Europe finally shifted the basis of commercial and political power from the Mediterranean and Indian Ocean to Atlantic Europe.

1. Pirenne, Henri, *Medieval Cities* (New York 1925), p. 10.
2. Leeds, E. T., 'Denmark and Early England', *Antiquaries Journal* 25 (1946), pp. 26–35.
3. Hornell, J., *Water Transport* (Cambridge 1946), p. 141; Arenhold, L., 'Nydam Boat at Kiel', *Mariners' Mirror* 4 (1914), p. 183.
4. Phillips, C. A., 'The Excavation of the Sutton Hoo Ship Burial', *Antiquaries Journal* 20 (1940), pp. 149–202; Stenton, F. M., *Anglo-Saxon England* (Oxford 1963), pp. 50–3.
5. Lethbridge, T. C., *Merlin's Island* (London 1948).
7. Pirenne, op. cit., ch. 2. See also East, W. Gordon, *An Historical Geography of Europe* (London 1966).

7. Brønsted, J., *The Vikings* (Harmondsworth 1960), p. 25.
8. Ingstad, Helge, *Land under the Pole Star* (London 1966); Marcus, G. J., 'The Greenland Trade Route', *Economic History Review* **7** (1954–5).
9. Tonning, O., *Commerce and Trade on the North Atlantic 850 A.D. to 1350 A.D.* (Minneapolis 1936).
10. Lewis, A. R., *The Northern Seas* (Princeton, N.J. 1958).
11. Williams, D. T., 'Medieval Foreign Trade' in Darby, H. C., (Ed.), *Historical Geography of England Before 1880* (Cambridge 1936), pp. 266–82. Also Moor, Sir Allan, 'Rig in Northern Europe', *Mariners' Mirror* **42** (1956).
12. Lewis, B., *The Arabs in History* (London 1958).
13. Mallett, M. E., *The Florentine Galleys in the Fifteenth Century* (Oxford 1967).
14. Hourani, G., *Arab Seafaring* (London 1963); Villiers, A., *The Indian Ocean* (London 1952).
15. Hirth, F., and Rockhill, W. W. (trans.), 'Chau Ju-Kua His Work on the Chinese and Arab Trade in the Twelfth and Thirteenth Centuries entitled, *Chu fan Chi*' (St Petersburg 1912). Also Rockhill, W. W., 'Notes on the Relation and Trade of China with the Eastern Archipelago and Courts of the Indian Ocean during the Fourteenth Century', *Toung Pao* **15** (1914) and **16** (1915).
16. Mookerji, R., *Indian Shipping* (London 1957); Wheatley, P., *The Golden Khersonese* (Oxford 1961); Van Leur, J. C., *Indonesian Trade and Society* (The Hague 1955).
17. Hoyle, B. S., 'Early port developments in East Africa', *Geografie* (March–April 1967), p. 16.
18. Latham, R. E. (trans.), *The Travels of Marco Polo* (Harmondsworth 1959), pp. 213–15.
19. Worcester, G. R. G., *Sail and Sweep in China* (HMSO 1966).
20. Hirth and Rockhill, op. cit.
21. Brendt, R. M., and Brendt, C. H., *Arnhem Land, its History and its People* (Melbourne 1954). Also Nelson, J. G., 'Pre-European Trade Between Australia, Indonesia and the Asiatic Mainland', *Canadian Geographer* **5** (1961).
22. de Beuclair, I., 'Some Ancient beads of Yap and Palau', *Journal of the Polynesian Society* **72** (1963), pp. 1–10.
23. Couper, A. D., 'Indigenous Trading in Fiji and Tonga: A Study of Changing Patterns', *New Zealand Geographer* **24** (1968), pp. 50–60.
24. Golson, J. (Ed.), *Polynesian Navigation, A Symposium* (Wellington 1962) and Fris, H. R. (Ed.), *The Pacific Basin* (American Geographical Society 1967). Both include extensive bibliographies on Pacific maritime trade and exploration.
25. Taylor, E. G. R., *The Haven-Finding Art* (London 1965).

3

BRITISH SHIPPING AND ECONOMIC GROWTH

In the mid-sixteenth century Britain possessed very few merchant vessels suitable for overseas voyages. Some small craft were sailing out of Bristol on the wine trade to Spain and Portugal, others were engaged in the wool and timber trades of the North Sea and Baltic, and occasionally voyages were made as far as Iceland. Apart from these there was little native-owned commercial shipping, and even on the coal trade from the Tyne to London there were only two English-owned vessels.[1]

The dominant nation in northern seas at this time was the Dutch, who, during the sixteenth century, had freed themselves from Spanish tutelage and had captured much of the trade of the Hanse. They developed a type of vessel well suited economically to sixteenth-century trade in the region. The Dutch fluytschip was a wide-bottomed vessel with a sharp tumble home and a simple sail arrangement. These features allowed maximum cargo space, and at the same time gave the lowest possible tonnage measurements for the payment of Sound Tolls to enter the Baltic, and also allowed the minimum number of crew. They were not fast craft and generally they carried no armaments; they were strictly low-cost commercial carriers designed for hauling grain, meal, salt, fish, timber and coal. Earlier medieval craft, although often larger than the Dutch fluytschips, were well armed and, being partly propelled by oars, required relatively large crews; moreover, merchants commonly travelled with their goods on these traditional medieval vessels so that the proportion of space on board for cargo was less than for the smaller fluyts.

The Dutch also traded fish to Portugal in return for oriental products. In the course of time they bypassed the Portuguese and entered the Indian Ocean to trade direct with south-east Asia. A base was established at Table Bay and Dutch vessels sailed in the zone of the westerlies almost to the coast of Australia before hauling northwards to Java. By infiltrating south-east Asia from this direction the Dutch drove the Portuguese out of the region into India. By the late sixteenth century they had gained a virtual monopoly of the spice trade to northern Europe.

British shipping developed in a significant way only in the late sixteenth century. At mid-century Britain possessed a reserve of excellent seamen in the coastal towns and fishing villages, a small force of naval vessels and a mere handful of merchantmen, some of which were privateers or quasi-privateers. In 1588, however, she was able to draw together experienced seamen and shipmasters from around the coast and assemble a sufficiently strong fleet of nimble well-armed small ships to defeat the Spanish Armada; the scene was then set for her ultimate emergence not only as mistress of the seas but as the primary trading nation of the world.[2]

The rise of British shipping represents, and was partially responsible for, a revolutionary advance in the British economy from being predominantly agricultural to one with a strong commercial basis. This was reflected in the founding by London merchants of overseas trading ventures, such as the Eastland Company in 1579, the Levant Company in 1581 and the Africa Company in 1588. During this period much bullion from Portuguese mines in Brazil came into English hands as English ships traded home-produced woollens, worsteds and serges to English agents established in Portugal.

One of the main keys to the expansion of British commerce from the sixteenth century onwards was the existence of a large merchant fleet, supported by a strong navy, which took advantage of the discoveries and the advances in cartography and navigation as well as Britain's favourable geographical position in a world turning increasingly to seaborne trade. By 1700 there were between 5,000 and 6,000 British-owned merchant vessels carrying coal and other cargoes in the coastal, continental, and overseas trades. At the same time British seamen numbered about 100,000; this constituting the largest non-agricultural sector of the working population.[3] These developments gave a new spatial dimension to the British economy which facilitated its subsequent industrial growth.

The maritime advance was, of course, aided by the Navigation Acts. But for an adequate explanation of the initial expansion in shipping we must look to the domestic coal trade; and especially to the sea-link between the coal resources of north-east England and the English capital. This trade carried so many other activities along with it that it undoubtedly represents one of the most important components in the British 'take-off' into rapid economic growth.

The domestic coal trade

By the sixteenth century most of the populated areas of Europe were suffering from an acute shortage of timber. Deforestation around the towns; the primitive nature of land transport; and the competing demands on the remaining accessible forests for naval purposes, building materials, charcoal and fuel, raised the price of timber. The securing of winter fuel supplies for the growing centres of population was particularly problematic.

Britain was fortunate in having coal which outcropped on the banks of the navigable waterway of the Tyne. It could be mined relatively easily to the west of Newcastle, carried downstream by keels, loaded on seagoing vessels, and distributed coastwise—especially to London. Until the late sixteenth century most of this coal was shipped by Dutch and other foreign vessels, but by 1700 there were over 1,000 British-owned ships engaged in the coal trade between the Tyne and London. The foreign tonnage had been effectively driven out partly through a tax imposed on merchants who shipped coal on ships other than British owned.[4]

It is likely that many of the coal vessels under the British flag in the latter half of the sixteenth century were still fluytschips acquired from Holland. However, in order to ensure regular supplies of coal it was necessary for Britain to build new craft suitable for the carriage of this heavy and bulky commodity, the scantlings of contemporary English ships employed in the carriage of wine and cloth being far too light for the coal trade. The east coast was also particularly hazardous for navigation. Easterly gales, in summer as well as winter, drove many ships ashore; there were few off-shore shelters and safe storm anchorages; the treacherous Goodwin Sands were inadequately marked with beacons; river entrances were generally unimproved; fog was ubiquitous; and enemy and pirate vessels frequently were active. These conditions called for vessels strongly constructed and easily manœuvred.

The brigs that evolved in the east coast coal trade were built with stems of solid baulks of oak and broad keels which allowed them safely to ground, and they were designed to facilitate easy loading and discharging of cargo. Such vessels were particularly nimble in restricted conditions and were easily handled by a small crew. These were the commendable features which Captain Cook recognised when he adopted a collier brig for his first voyage of discovery which began in 1768.

Most of the colliers were built on the Tyne, at Whitby, Ipswich, Yarmouth or London. They stimulated a new era in shipbuilding which drew capital and labour not only to shipyards but also to iron and other metal industries and to the manufacture of sails and cordage. Furthermore, the east coast coal trade became the main nursery of British seamen for foreign-going voyages, and the seamen employed in it represented a vital naval reserve.

As the coal which outcropped along the banks of the Tyne became exhausted mining proceeded further inland. But always it had to connect with the collier fleets. Wagon-ways, the early precursors of the railway, led into County Durham and Northumberland and these connected the new inland mines to riverside loading staithes, from which fleets of heavy keels carried the coal downstream to the ships. By the mid-eighteenth century coal-mining in County Durham had moved eastwards on to the concealed coalfield, and steam-driven drainage equipment had been developed to make this possible. The juxtaposition of wagon-ways and steam engines in north-east England signified future transport developments in the form of mobile steam engines driven on railways.

Probably even more important than all these growth factors associated with the coal trade was the significance which it had for the expansion of London, and all which this implies in the commercial dominance of Britain in the world of the eighteenth and nineteenth centuries. The colliers sailing to London grew in size from an average of 73 tons in 1606 to 248 tons in 1701. From the Tyne, and also the south Wales coalfields, about 8,000 tons of coal per annum were shipped by sea to the capital during the late sixteenth century. A century later this had risen to 400,000 tons, and by the early eighteenth century the quantity reached about 700,000. This was London's lifeline. The colliers brought cheap fuel right to the Pool, at the heart of the city, and their berths lined the river between Blackwall and Woolwich. The regularity of sea-coal deliveries encouraged industries in London and provided domestic fuel; this, and the employment generated by the increasing port activities, must have contributed enormously to the support of London's population which grew from 200,000 to 700,000 between the beginning of the seventeenth century and the mid-eighteenth century. It was then the largest city in Europe and had a population greater than all the other urban areas in England put together.[5] By the beginning of the nineteenth century London's population exceeded one million. The city, at various times, attracted immigrants not only from other parts of Britain but from all over Europe; they included Huguenot and Jewish merchants who contributed to its financial power.

The sheer size of London gave it an element of self-sustained economic growth, and the merchants who were serving the needs of this immense population must have accumulated capital rapidly. Much of this was invested in overseas commerce. The growing demand for luxury goods on the part of London's middle and upper classes in turn attracted most of the overseas shipping from Europe, the Far East and the West Indies. This contributed to the establishment of London as the primary port of Europe, and London's coffee houses pulled the centre of marine insurance away from Amsterdam.

The domestic coal trade of the seventeenth and eighteenth centuries thus represented a vital link in the new spatial organisation of the British economy. It vitalised shipbuilding and other industries, and also, by supporting London, helped make possible the concentration of financial capital employed in international commerce. This latter activity, the overseas trading venture, was the second essential component in British economic growth. It is to this we must turn in order to appreciate the full significance of shipping in the spatial extension of the British economy and, *pari passu*, its dominant role in the world during the nineteenth century.

Development of overseas trade
By the mid-seventeenth century the centre of maritime commerce had shifted from Spain and Portugal to northern Europe. This may be explained partially by differences between the domestic economies of the North Sea lands on the one hand and the economy of the Iberian lands on the other, which developed on the basis of their respective maritime activities. In brief, the commerce of Spain and Portugal was parasitic on colonial territories. By contrast the Netherlands and Britain, in order to acquire luxury goods and bullion, were stimulated to sell shipping services to other nations and also to produce manufactured articles for exchange. As the domestic industries of the latter countries expanded, the English, in particular, sought new trading outlets overseas. They were unable to break the Dutch monopoly in south-east Asia, but they defeated the Portuguese in Indian waters and through the East India Company laid the foundations of an Eastern colonial empire. They gradually eroded the commercial functions of Amsterdam; and when this city suffered in the Napoleonic wars London was confirmed as the centre of world trade.

The new trading patterns centred on Britain were very different from traditional maritime linkages. At this time Britain, which was rapidly industrialising, commanded a network of shipping links and a powerful navy to protect it—naval bases having been established at strategic positions. Manufactured goods were carried outwards from Britain and raw materials flowed inwards. Whereas earlier trade relations had merely emphasised the indigenous resources and methods of production in overseas countries, as in the silk and spice trades, the new trading network gave rise to basic changes in land use and to major shifts in population; thus ultimately they altered completely the balance of trade. By the nineteenth century it was no longer oriental luxuries that came via the Cape of Good Hope, but bulky raw cotton, jute, wool, grain and rice. The high-value goods (including cotton textiles for India) now flowed in the opposite direction as the products of British manufacturing industry. British shipping, and commerce, thereby changed much of the world and knit it into greater economic interdependence than ever before. It also gave to many overseas regions their distinctive characters in respect of port and hinterland activities.

Associated with the world circulation of goods carried on British ships there also flowed through the City of London much of the world capital used in international commerce. Already by the seventeenth century the slow capital turnover, from investments in expensive ships which were making longer voyages with valuable exports, necessitated the formation of companies commanding considerable financial resources. These companies began to finance merchants with goods moving on the seas between buying and selling areas. The insurance of the ships and cargoes by London underwriters, the basis of which had been established at Lloyd's Coffee House in 1687, dominated world marine insurance; and the Bank of England, founded in London in 1694, took over the financing of most of the cargo consignments moving in international trade. The related tasks of matching ships to cargoes, which had for centuries taken place at London's Virginia and Baltic Coffee House, was in the early twentieth century concentrated in the Baltic Exchange, at which almost all international shipping fixtures were thereafter made.

The world-wide changes which came about in land use and population during the building of British maritime institutions, and Britain's Empire, may best be illustrated by reference to one or two of the principal trading regions. Between 1638 and 1788 ships from London, Bristol and Liverpool participated in an Atlantic triangular trade in which metal and glass manufactured goods were carried to West Africa, and slaves from there to the West Indies, South America, and North America, from all of which cargoes were obtained for Europe. Some of the vessels returned directly from the West Indies with sugar and rum, while others loaded cotton in the Mississippi, tobacco in Virginia, or rice at Charleston. Through the slave trade, conducted by Britain and the ships of several nations of north-western Europe, the racial composition of the American continent was influenced for all time. In the sixteenth century, for example, slaving vessels carried about one million Negroes from Africa; in the seventeenth century three million; and in the eighteenth century seven million were transported.

Further north in the American continent other British vessels were loading timber and masts in New England; and timber, furs, and skins in Canada. But even in the eighteenth century the number of trading voyages which a ship could make in any given time was limited. On the trade direct to the West Indies for example a ship could achieve only two or three round trips per annum; while a round voyage to India would take over a year. On the other hand, with the increased availability of cargoes, especially from the British colonies, bigger ships could be employed. Productivity also improved: Davis notes that in the trade between London and America the average ratio of crew to tonnage in 1686 was one man per 9·8 tons. By 1766 the average had increased to one man per 15·6 tons.[6] The elimination of much privateering, which previously necessitated the arming and over-manning of merchant vessels, was partly responsible for this. Ship-carrying performance, however, was limited by slow methods of handling cargo and by incomplete knowledge of wind systems on which efficient ocean voyaging depended.

An East India Company ship of 1,000 tons, for example, might spend weeks beating out of the English Channel. Her course would then be set, via Madeira, to reach the north-east trade winds, and under these she would proceed west of south until the limits of the trades were reached. Such a ship would often experience further difficulty in passing through the zone of variables in order to reach the south-east trades, but having picked these up would sail as close as possible to the wind as she approached the coast of Brazil. An India-bound vessel would then maintain this course until reaching latitude 30° to 35°S. The westerly winds would then be met, and with these she would run down her easting to Table Bay.[7]

The arrival of an East Indiaman in the Indian Ocean was often timed to coincide with the south-west monsoon, especially when bound for a Bay of Bengal port. Ships for China and Australia would follow a similar route to Cape Town, thereafter maintaining a rhumb-line track eastwards until the coast of Australia was reached. In the first quarter of the nineteenth century even the best of ships would take over 140 days on the passage from Britain to the colony of New South Wales.

There was a remarkable improvement in sailing performance on the Australian run during the second half of the nineteenth century. This doubtless arose from such factors as: the publication of wind and current charts by Lieutenant Maury of the US Navy; the availability of the chronometer; and the wider adoption of the fine-lined hull introduced by American clippers during the Californian gold rushes. Maury's charts showed seamen how to make the best use of seasonal and prevailing winds and currents on great circle routes and the use of the chronometer allowed the accuracy of position finding necessary for these sailing practices.⁸

American-owned clippers, and those purchased from New England by British owners, entered the Australian trade in the 1850s. They made the normal wind-determined sweep of the Atlantic, but from about the latitude of Tristan da Cunha they followed a new route on a full or composite great circle track to Melbourne, often going as far south as latitude 60° (Fig. 1). Passages from Britain were, as a result, reduced from well over 100 days to 70 days, and even less. By bringing it closer to Europe emigration to Australia eased significantly. Between 1787 and 1840 the cumbersome ships which followed the long rhumb-line route from Europe to Australia via Cape Town carried about 180,000 convicts. But in the same period no more than 15,000 free migrants entered Australia, for few people would voluntarily choose to settle in a country almost five months' voyage away. The coincidence of improvements in sailing ship design; the new knowledge of the oceans; the discovery of gold in Australia; and the introduction of assisted passages, released Australia from what has aptly been termed the tyranny of distance.⁹ The fast passages of the Baltimore and British-built clippers helped increase the Australian population from half a million in 1851 to over one million by 1861.

The movement of people from Europe to North America was on an even bigger scale than that from Europe to Australia. The existence of cheap passages across the Atlantic, especially on vessels engaged on the timber trade returning to North America in ballast,¹⁰ served as the link between the 'push' being given to much of Europe's agricultural population, partly as a result of cheap imported grain and wool, and the 'pull' of the opportunities in the New World. Between 1800 and 1875 more than seven and a half million people migrated to North America from the British Isles. This was undoubtedly a period of one of the greatest maritime migrations in history.

In the middle of the period the Navigation Acts were rescinded, and in the early 1850s the new American merchant fleet, with five million tons of sailing ships, almost equalled that of the British. But by then Britain's lead in industrial development, based on advancing technology and the exploitation of coal and iron resources, was making itself felt in the maritime field with the appearance of ocean-going steamships.

The advance to steam and iron ships

Britain was already an industrial country in the first quarter of the nineteenth century. The canals had by then opened the inland coalfields, and a steam tug had already functioned on the Forth and

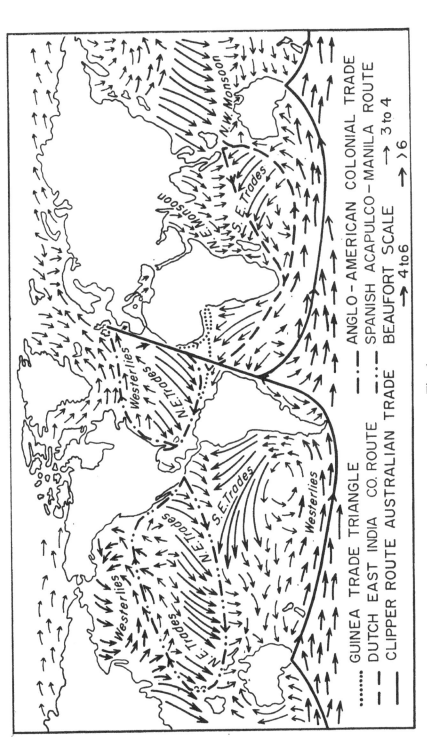

Fig. 1
World wind systems (January) and ocean routes of European sailing vessels.

Clyde Canal. A few years later combined sail and paddle steamers crossed the Atlantic; then in 1838 the first steamship built for this service, Brunel's *Great Western* of 2,300 tons displacement, came into operation. This was followed by the 3,678 ton *Great Britain* in 1846, also built by Brunel and fortunately preserved as the first iron-hulled and propeller-driven ship ever to cross the oceans. Brunel's mammoth *Great Eastern* of 22,600 tons was not commercially successful; but it was now clear that the constraint which the use of timber had imposed on the size of ships was removed, and henceforth only the demand for cargoes and port factors need be considered.

For all their speed the sailing clippers could not compete with the improvements in steamships. To achieve their phenomenal reductions in time on ocean passages the streamlined clippers had sacrificed cargo space, and in order to minimise draught in the interest of speed they retained wooden hulls into the age of iron. The engines of steamships steadily improved in efficiency and reliability, scheduled services were introduced, and iron construction allowed much larger ships to be built. Ocean passages were now no longer governed solely by the wind systems, and time in port was also reduced with the application of steam to mechanical equipment for cargo handling. Finally, when the Suez Canal opened in 1869, and the advantages of the high latitude great circle routes diminished, the age of the sailing ship was almost over. They survived in significant numbers only on the wheat trade from Australia and western Canada, the nitrate trade from Chile, and in the carriage of coal outwards from the British Isles, up until the First World War.

The development of steam at sea coincided with the spread of steam locomotion on land. New agricultural areas were opened by the railways and vast quantities of raw materials and foodstuffs flowed to the ports and were carried onwards mainly on British ships to Europe. British shipping tonnage rose in this era of the beginnings of steam from five million tons in the mid-nineteenth century to ten million tons by the end of the century. The steam age meant also that the railways and bunkering stations overseas required vast supplies of coal. As a result British tramp ships were certain of profitable outward cargoes, and could return home with grain and raw materials accepted at relatively low, but profitable, freight rates.

The arrival of cheap grain at the ports of Europe reduced food prices, but it also gave rise to unrest amongst many of the rural populations extending from Imperial Russia to Ireland. The great migrations across the Atlantic continued under this impact. The period likewise witnessed many changes in the British population, both in numbers and relative distribution. In 1800 the population of Britain was about 10 million, of whom 70% were rural; in 1900 the population had reached 40 million, of whom only 30% were to be found in rural areas.

By the last quarter of the nineteenth century meat and dairy produce were also arriving in Britain from Australia and New Zealand. Up until this time the climate of the tropics and distance had presented barriers to the transport of fresh produce. The development of refrigerated steamships released the tremendous meat resources of Australia and New Zealand which had hitherto been exported only in limited quantities in salted form or rendered down as tallow. This contributed to a further growth in population, for by making more protein available to the people of the densely populated industrial areas mortality rates declined. Meat consumption rose steadily in nineteenth-century Britain, and by the first quarter of the twentieth century half the meat eaten in this country was brought from overseas by refrigerated ships.

Just as the tramp ships and railways affected the pattern of agriculture in the prairies and pampas so the establishment of regular liner shipping to carry fresh products influenced the direction of future developments in Australia and New Zealand. The climate of the latter country, for example, was ideally suited to the outside grazing of livestock throughout the year, and with the introduction of refrigerated ships New Zealand became the overseas farm of Britain. Exotic grasses were introduced and fertilisers imported. The result was that within a few decades the landscape of New Zealand was transformed and the economy became highly dependent on a country 13,000 miles away.

The efficiency of shipping was also enormously improved in the last quarter of the nineteenth century by the invention of the electric telegraph. Information on world surpluses and shortfalls could now be communicated between owners and shipping agents throughout the world and the ships suitably directed. The communications basis for a world economy was now complete; a centralised freight market came to be controlled from the Baltic Exchange, while London bankers financed most of the cargoes being bought and sold between overseas merchants.

Accurate figures do not exist to reflect the phenomenal increase in the circulation of manufactured goods and primary products throughout the world during this period. However, the expansion of the tonnage of the world fleet of merchant ships indicates this increase; it rose from seven million tons in 1850 to twenty-nine million tons in 1900, in which period Britain owned almost half of the tonnage and carried more than half of the world's seaborne trade.

Industrialisation also proceeded in other countries, particularly in Germany, the United States and Japan; and these too expanded their merchant fleets. Britain was, however, dominant, and she retained advantages so long as the age of coal fuel and power lasted in transport and industry. Somewhere in the region of eighty million tons of coal continued to be shipped annually from British ports during the 1900s, and this provided the outward cargoes for the thousands of British tramp ships. British passenger shipping also thrived at this time, for the earlier migrations across the Atlantic had left a legacy of cultural interaction. This not only pulled others to join relatives in North America but it established a tradition of vacational visits to Europe. This is reflected in the building of the giant transatlantic passenger liners such as the *Mauretania*, of almost 32,000 gross tons and capable of over 26 knots, which was launched ni 1907.

By the outbreak of the First World War the British merchant fleet stood at over eighteen million tons. Its nearest rival, Germany, owned almost five million tons and the United States possessed two million. But just as her merchant fleet had supported and reflected economic growth in Britain during the industrial revolution, so in the depression years which followed the First World War Britain's merchant shipping stagnated. The United States also suffered economic recession, which, amongst other things, adversely affected the British passenger-carrying trade. A system of American government subsidies, however, helped increase the tonnage of the US fleet.

There was also, following the First World War, a significant shift from coal to oil as a source of energy. The outward cargoes of coal on which the tramp ships relied were from then onwards steadily eroded. The new type of oil-fuel cargoes to be loaded in the United States and the Middle East required changes in maritime technology. On this occasion the British shipping industry was slow to respond. Norway, on the other hand, started to emerge more definitely as an advanced nation greatly helped by the earnings of tankers engaged in the oil trade, and between the First World War and 1939 the Norwegian fleet doubled in size. Britain, while remaining the principal maritime nation, commanded in 1939 only a little over one quarter of the total world tonnage of merchant ships.[11]

The remarkable growth of British industry and trade in the eighteenth and nineteenth centuries can be attributed, in no small measure, to the cheap carriage of imports and exports by British shipping, and to the creation of new raw material bases and markets overseas. The invisible earnings from selling shipping services to other countries gave to Britain a favourable balance of payments, as did the dominance in international marine insurance, and the credit and brokerage services provided by the City of London to world shippers. This spatial extension of the British economy to global dimensions in the nineteenth century was made possible first by advanced shipbuilding and engineering technology and second by the efficient and cheap operation of the world's largest fleet of merchant ships. The role of the coastal coal trade in laying the basis for the industrial revolution and British maritime supremacy should, however, be recognised.

In the process of controlling and guiding the international circulation of goods, British shipping and commerce changed the geography of many parts of the world. Vast, almost monocultural, areas of raw material and food production appeared in many regions; great numbers of people were conveyed by sea to create new communities in colonial areas, and to supply labour for expanding plantation and mining economies. The plural societies in many parts of the earth owe their origins to this rearrangement of the world economy by British maritime transport during the eighteenth and nineteenth centuries.

1. Nef, J. U., *The Rise of the British Coal Trade* (London 1932), p. 72. See also Smith, R., *Sea Coal for London* (London 1961).
2. Davis, Ralph, *The Rise of the English Shipping Industry In the Seventeenth and Eighteenth Centuries* (London 1962).
3. Hobsbawn, E. J., *Industry and Empire* (Harmondsworth 1969), p. 25.
4. Nef, op. cit., p. 172.
5. Smailes, A. E., *The Geography of Towns* (London 1953), p. 22.
6. Davis, op. cit. See also North, D. C., 'Sources of Productivity Change in Ocean Shipping 1600–1850', *Journal of Political Economy* 76 (1968), pp. 953–65.
7. Parkinson, C. N. (Ed.), *Trade Winds* (London 1948). Provides an account by authoritative authors of the British maritime trades between 1793 and 1815.
8. Maury, M. F. (Leighly, J., Ed.), *The Physical Geography of the Sea, and its meteorology* (Cambridge, Mass. 1963).
9. Blainey, G., *The Tyranny of Distance* (Melbourne 1966).
10. Guillet, E., *The Great Migration* (Toronto 1963). See also Greenwood, Basil, *The Great Migration crossing the Atlantic under sail* (National Maritime Museum, HMSO 1968).
11. Rochdale Report. For details of interwar figures and bibliography on shipping see *Committee of Inquiry into Shipping* (HMSO, Cmnd. 4337, London 1970).

4

WORLD SHIPPING ROUTES

The flow maps used in geographical texts to depict the movement and scale of seaborne trade are helpful in providing a synoptic view of economic interaction in the world. This is sufficient for many purposes. But the actual pattern of ship movement is also important since it relates to man's use of the oceans as a highway; a use which has frequently changed with technological and political conditions.

Flow patterns also reflect a popular perception of the oceans as a medium unimpeded by physical or political obstacles; this is something of a distortion. It is true that the sea presents less frictional resistance to transport than does the land, and the high density and low viscosity of sea water allows the support and ease of movement of vehicles of enormous tonnages, but the seas are in no sense free of impediments. Shipping movements have been influenced at various times and at various stages of technology by physical conditions, the need for portages and canals, and the politico-legal divisions of coastal and even ocean areas. At any one time the choice of a route has been a matter of balance between speed, safety, and costs; in this the owner and master have taken into account the capability of the vessel in relation to the physical, economic, and political factors of the operating conditions.

Physical factors
The surface of the sea is seldom calm. In some areas at all times and in other areas at certain times of the year waves of great destructive force may be encountered. Zones of severe wave activity exist in the Atlantic and north Pacific during winter, especially between latitudes 50° and 60°N; corresponding zones are to be found in the southern ocean, in the Indian Ocean during the south-west monsoon season,

and in the tropical south-west Pacific, West Indies and China Sea with the passage of a tropical storm. Such sea conditions confronting a ship comprise a large number of wave trains of different height, length and frequency. Depending on the fetch, duration, and strength of the wind they build up to high moving ridges, often extending for over 15 m (about 50 ft) from crest to trough, and present formidable obstacles which result in a reduction in speed, alteration of course, and possibly damage to vessels.

Damage may occur in several ways; through the slamming action of the bow as the ship rises and falls head-on to the waves, through severe vibration as the propeller is lifted clear of the water, and when solid water or 'green seas' are shipped when the vessel is labouring in a seaway or rolling to windward. Paradoxically, rough seas may present greater dangers to power-driven ships than they did to sailing craft. The latter tended to move with the wind and sea, the former has the power to challenge the sea head on. This means power-driven ships have to be nursed in rough seas. Failure in this regard still too often results in the loss of a ship,* more frequently it means damage to the hull, superstructure, or cargo. With the rising costs of repairs, cargo claims, and insurance premiums, shipowners are anxious to avoid any additional expenditure arising from this, and they are becoming more than ever aware of the environmental hazards which may accompany damage to ships carrying oil or chemicals.

Many modern vessels which operate in areas of rough seas are expected to run to schedule, and they may carry containers and other deck cargoes in a vulnerable position. Time for these vessels means money, but a master must balance the risks involved in maintaining his ship's course and speed against loss of time by reducing engine revolutions or altering course. Until very recently a master had only limited information as to the best route to adopt to avoid high wave obstacles. It is now possible for him to receive, by radio-linked equipment, daily forecast maps of the sea conditions which show the wave contours for 24 hours ahead. The course can then be plotted each day through wave fields with the lowest gradients. Alternatively, a master may receive information from a shore-based meteorologist on the optimum route to adopt. By using these techniques of weather routeing a ship can follow the least time track to avoid or reduce delays, as well as to obviate the possibility of damage by wave action. Vessels on the north Atlantic trade can in winter save 24 hours or more on a single passage by weather routeing.[1]

Ice is another condition which presents hazards to shipping and may

Lloyd's Register Casualty Returns lists 108 vessels as lost through foundering in 1969.

involve alterations of course, and in some areas the temporary suspension of navigation. However, modern power-driven ships with strong hulls and accompanied by ice-breakers have reduced the constraint imposed by ice on navigation. The north-east passage between the Kara Sea and the Bering Strait across the Siberian seas can now be maintained for 160 days in a year. The open period may even be extended by using more craft like the nuclear-powered ice-breaker *Lenin* of 16,000 tons which can force its way through ice over 3 m (10 ft) thick. As well as trading to the ports of Siberia for timber and minerals ships have used this route as an alternative passage between western Europe and the Far East, for it reduces the distance even over the Suez Canal route by 2,000 miles.[2] The north-west passage across North America has also been shown to be feasible for shipping since the voyage of the specially strengthened tanker SS *Manhattan* in September 1969. Through its sheer bulk, and specially designed ramming bow, the 115,000 dwt *Manhattan* broke its way through all but the thickest ice. Bigger ships with more engine power than the *Manhattan* may, therefore, relatively easily make the passage. This has opened the possibility of mineral and oil shipments by sea from Alaska and the Arctic islands to the Atlantic all the year round, and it offers an alternative high-latitude route between the Atlantic and the Pacific should it ever be required, for the distance from London to Tokyo is about 8,600 miles by way of the Suez Canal, 14,600 via the Cape of Good Hope, and 8,000 miles by way of the north-west passage. But too much significance should not be put on reduced distance, for time can be lost on these ice routes. This may be due to waiting for ice-breakers and convoys and to reduced speeds in areas of ice packing. The costs are also raised above those on the longer routes due to charges for ice-breakers, ice insurance rates, and necessary ship strengthening.

Ice in the Gulf of Bothnia is still an impediment to navigation of small vessels and effectively cuts off the ports of Sweden and Finland which border that sea. But whereas the southern Baltic was closed by ice for six months in the year during the days of sail a channel can now be cut by ice-breakers through the ice to the main ports during most of the winter. Similarly, specially strengthened ships trading on the north Atlantic can reach Montreal throughout the depth of winter, although the St Lawrence Seaway into the Great Lakes cannot be kept open. The Strait of Belle Isle is also closed by ice, and some north Atlantic shipping must divert to the open water ports of St Johns (Newfoundland), Halifax (Nova Scotia) and St John (New Brunswick).

Even in the open sea ships must frequently divert from shortest distance routes because of ice hazards. Icebergs calved in summer from the glaciers of west Greenland drift southwards with the current of the Davis Strait until they are brought to a halt by pack ice in the winter. They are freed again in the spring and many of them eventually reach the shipping lanes. However, ice patrols organised by US coastguards, using air and surface craft and the radar which is fitted on most vessels, have reduced these dangers. Radar has also reduced the possibility of collision and grounding in fog; but so-called radar-assisted collisions occur, due largely to human errors in interpretation, and fog still slows ships and adds to their difficulties in narrow congested areas, such as the Strait of Dover where it can be expected on 30 days in the year.

Technological advance has clearly removed many of the physical constraints on shipping, but it has also increased the effects of others. One of the most pronounced advances in shipping is the greater carrying capacities of modern vessels, with the corollary of deeper draughts. In order to obtain the economies to be derived from size, especially in the carriage of oil, many ships of from 200,000 to over 300,000 tons, with loaded draughts of 21·34 to 24·3 m (70 to over 80 ft), were already operating in 1969. Such vessels cannot enter the Baltic fully loaded, nor pass through the Strait of Malacca or Singapore Strait, nor use any of the international waterways, and the largest class are unable to reach North Sea ports through the Strait of Dover.

The trend to deeper draughts has brought about a revaluation of routes and channels and new patterns of movement. Large ships now focus on a few deep-water locations from which transhipments are made by pipeline or smaller vessels. The giant ship has also brought about the need for more detailed knowledge of the marine environment; particularly information on short-term changes in bottom contours due to shifting sand ridges, more precise off-shore tidal data, more details on the likely effects of meteorological conditions on the rise and fall of the tide, and especially on the phenomenon of negative surge when the predicted level of the tide can be depressed by as much as 1·83 m (6 ft).[3]

The tidal rise is particularly important for deep-draughted ships. On entering coastal waters these vessels may 'squat' and increase their draught by over 0·91 m (3 ft); they may also encounter wrecks which tend to shift position, and in the much traversed North Sea there are over 5,000 known wrecks. Gas, oil, or chemical-carrying ships navigating in restricted conditions, in close proximity to land and areas of fishing activities, can clearly present hazards. From the operator's point of view, however, every additional foot of loading can mean an extra 4,000 tons of cargo on vessels of 200,000 to 300,000 dwt. The full economies of these ships and the protection of the environment can be reconciled therefore only by accurate knowledge of physical conditions and strict traffic control along the channels.

International waterways

The principal canals have for long acted as the foci of international sea transport. When the Suez Canal opened in 1869, for example, it immediately attracted steamships, for it reduced the distance between London and Bombay by over 4,000 miles, and brought Australia about 1,000 miles closer to Britain. It put the Mediterranean ports back on a channel of international trade, drew steamship traffic away from the Cape of Good Hope route, and revived ancient trading places such as Aden and Colombo to act as bunkering ports.

The Suez Canal route was so important for international shipping that shipowners for almost a century refrained from building ships beyond its draught limits, and these increased by less that 0·91 m (3 ft) between 1908 and 1958. The closure of the Suez Canal during the Arab–Israeli conflict of 1956–7 saw the first major departure from the canal route since the nineteenth century. The diversion of shipping to long hauls around the Cape at this time both allowed and made necessary the building of ships beyond the canal limits. At the termination of hostilities the Suez Canal authorities made immediate efforts to recapture tonnage and to attract the deeper-draughted ships then building. Between 1958 and 1967 they succeeded in increasing the permissible draught in the canal from 10·5 to 11·58 m (34½ to 38 ft) and embarked on a further deepening programme.

The outbreak of the second major conflict in the Middle East during 1967 closed the canal once again and halted the dredging and widening operations. At that time the canal had succeeded in attracting back most of the shipping and had gained from a steady build-up in the oil trade. Of the 242 million tons of cargo which passed through the canal during 1966–7 over 70% was oil. Since 1967 ships have again increased in size to meet the needs of the Cape route. By January 1970 about 40% of world oil-tanker tonnage comprised ships with draughts of over 13·11 m (43 ft), and tankers with draughts exceeding 18·29 m (60 ft) accounted for three-quarters of all the tanker tonnage on order in the world at that time.

The Suez Canal could, of course, be deepened to take tankers of over 200,000 tons with draughts of more than 18·29 m. But despite the relative ease of civil engineering in the flat sandy isthmus the capital expenditure involved in carrying out this task would be in the order of £300 to £400 million. This sum would in all probability have to be recovered from canal dues. It should be noted in this connection that the cost of using the canal, at 1966 levels of dues, plus the cost of ship's time on canal transit and waiting for convoys, etc., almost equals the cost of the additional mileage for a 200,000-ton vessel going around the Cape of Good Hope with cargoes for Europe. This close comparison means that any significant increase in dues could confirm the Cape route for the now common size of ship above 200,000 tons. If in the face of these higher charges sufficient tanker tonnage is not drawn back to the canal route, and the canal dues are further raised to compensate for the loss of tankers, then some other types of vessels may also be permanently diverted and the financial position of the canal could be rendered critical. Taking account of these factors alone, and leaving aside political and military considerations, it is clear that the future influence of the Suez Canal on the pattern of world shipping has been radically reduced in recent years.

The Panama Canal by contrast still exercises a strong influence over ship design and the routeing of vessels. The maximum draught in this waterway is 12·19 m (40 ft), although in the dry season between January and April the lowered water level of Lake Gatun may reduce the permissible draught to 10·67 m (35 ft) or less. New bulk carriers of about 67,000 dwt trading regularly on the US Gulf–Japan trade have been designed to Panama Canal limits (Panimax ships), and so also have some container vessels; so far this canal has been bypassed only to a limited extent. Some very large ships carrying coal from Hampton Roads (US) to Japan, for example, have followed the Cape of Good Hope route. But there is no doubt that with the economies to be derived from giant ships, and the possibility of sending containers by rail across the 'land bridge' of North America *en route* for Japan, there could in the future be a relative lessening of the pull of the Panama Canal for long-distance trades.

A more fundamental difference between the Panama and Suez Canals lies in the commercial and strategic significance which the Panama Canal has for the United States. The main flows of shipping through the Panama Canal have been between the east and west coasts of the United States, and between the Far East and America's eastern seaboard. In 1967 about 87 million tons of cargo passed through the canal, less than one-quarter of which was oil. About 60% of the total cargo moved from the Atlantic to the Pacific and of this one-third originated in United States ports. Many of the cargoes were carried by ships of 12,000 to 30,000 tons suited to the transatlantic and inter-coastal trade of the American continent.[4] The amount of cargo likely to be generated on these trades in the future will ensure that there is a steady build-up of traffic to compensate for any losses through diversion of giant ships. However, when transits reach the level of about 20,000 ships per annum then delays will occur. In 1968 there were 14,000 transits; this situation could therefore arise in the 1980s, although there may, in the interim period, be a radical reduction in the numbers of conventional cargo liners in favour of smaller fleets of container vessels, thus lessening the pressure on the canal.

For reasons of ship size, possible saturation of the Panama Canal, and military factors, the United States Interoceanic Canal Study Commission have recommended a new transocean canal be built. The present Panama Canal cannot easily be enlarged without enormous costs and a long period of disruption of interoceanic shipping, for there is an almost unalterable system of massive locks whichr aise and lower ships to and from Lake Gatun 26·21 m (86 ft) above sea level, and there are intractable problems of water supply in the lakes. The Commission therefore recommends the building of a parallel sea-level canal across Panama, about 10 miles from the present canal, to take ships of up to 150,000 dwt.

The only other international waterway with any significance as an influence on the pattern of movement of modern shipping is the Kiel Canal. This crosses an ancient portage zone between the Baltic and the North Sea. The canal was opened in 1893 to obviate the haul around the 250-mile-long Jutland peninsula; it can now accommodate vessels drawing 9·45 m (31 ft).

As with the Panama Canal constructed by the United States the Kiel Canal was built by Germany partly for strategic purposes. It has since been of particular commercial importance for the trade between Hamburg, Bremen and the Baltic ports. In 1969 the principal users, by ship tonnage, were West Germany 27%, Poland 12%, Finland 9%, USSR 6%, Holland 5%. In total about 60 million tons of cargo was carried via the canal in 1969 and over 70% of the vessels making the transit were of less than 500 g.r.t.

The Kiel Canal has lost some traffic in recent years due to the growth in size of ships on the Swedish iron-ore trades. Also, increased canal dues have caused Finnish shipowners to divert vessels around the Jutland peninsula. With faster ships of over 16 knots it is no longer worth while incurring canal delays and canal dues on voyages between Finland and Britain.[5] But linking as it does the North Sea and the Baltic, where small ship and barge traffic is active, and lying in proximity to EEC, EFTA and COMECON trading areas, the Kiel Canal must continue to be of importance for the trade of north-west Europe.

Political and legal factors
The doctrine of *mare liberum* has given rise to the concept of complete freedom of the seas for vessels sailing outside recognised

territorial waters. Were this true then political and legislative factors would exert little influence on patterns of ship movement; in fact they exert considerable influence. The oceans of the world are, of course, free inasmuch as any nation may operate ships on them without the permission of other nations. The only interference with the free passage of shipping on the high seas which is generally sanctioned is the right to stop vessels suspected of slave trading or piracy, but ships may also be stopped under international law in order to verify their flags, in self-defence when there is a threat to peace (as in the action of the US Navy during the Cuba crisis), or where a blockade is in force (as in the action of the British Navy in preventing maritime trade with Rhodesia). This does not imply that masters or owners are otherwise free to move or conduct their vessels in any way they wish. They are always subject to the laws of the state under whose flag they sail, although in the case of 'flag of convenience' tonnage such laws are not very stringent. Ships are also governed by international regulations for preventing collision, and are prohibited from engaging in activities considered detrimental to coastal states and to other users of the sea. For example, it is prohibited under the amended 'Oil in Navigable Waters' Act (accepted as law by thirty-six nations in 1967) to discharge persistent oil into the sea within certain zones, and as a result oil tankers washing their cargo tanks on their way to loading ports have generally followed routes outside these zones. In the future the discharge of oil will be prohibited anywhere on the world's seas.

Freedom of the seas does mean that ships have rights to use international waterways and certain straits without hindrance. In the case of the Suez Canal it was explicitly stated under the Convention of Constantinople in 1888 that this international waterway would be free and open to the ships of any nation at peace and war without distinction. The right of passage is less explicitly stated for the Panama Canal. The Kiel Canal had imposed in the Treaty of Versailles the obligation to admit the transit of ships of all nations at peace with Germany. Important channels such as the Dardanelles, the Bosporus and the Strait of Gibraltar are equally, if not more, vital for access to major trading areas, and as such all vessels have the 'right of innocent passage', which means they have access provided they do not impinge on the security and rights of the adjacent coastal states. The St Lawrence Seaway passes through Canadian and US territories and is also referred to as an international waterway, but it has never, as a matter of law, been opened as of right to the vessels of all nations.[6]

The broad principles of freedom of the seas and the specific rights of coastal states to protect their own interests can obviously come into conflict. Israeli shipping, for example, in order to reach the Red Sea from their port of Eilat in the Gulf of Aqaba had, in the pre-1966 period, to traverse waters claimed by Egypt and Saudi Arabia. The latter countries questioned the innocent nature of Israeli shipping, a situation which contributed to the seizing of the Sinai Peninsula by Israel in order to command the Strait of Tiran and navigation in the Gulf of Aqaba.

The tendency in recent years has been for nations to extend their jurisdiction over wider areas of the adjacent seas for purposes of defence, customs and immigration, resource exploitation, and conservation. There are five zones radiating outwards from the coast in which varying degrees of control may be exercised. The first is internal waters, including estuaries and certain bays; the second is territorial waters which extend from the margins of the internal waters outwards for three miles or more. The laws of the state apply to shipping within these two zones. A less clearly defined 'contiguous zone' extends from the outer margins of the territorial waters. The extent of this third zone varies, but the distance from the outer margins of internal waters to the outer margins of the contiguous zone is generally accepted as not exceeding 12 miles, so that a state which claims 12 miles of territorial water would not have a recognisable right to a contiguous zone. In this latter zone customs and immigration controls may be enforced and primary rights of the nation regarding fisheries recognised. Finally, beyond this there is a broad, ill-defined, 'zone of diffusion' within which states may attempt fishery protection and may carry out naval manœuvres.[7]

In some regions of the world, such as Far Eastern waters, ships will deviate to avoid sections of these zones. Shipping is also tending to follow recommended routes laid down by international agreement as they pass through straits and channels and around headlands on which many vessels focus. The compulsory routeing of all ships in the English Channel, the Sound (between Denmark and Sweden), the straits around Japan, and other areas of high traffic density will undoubtedly be introduced in order to minimise collision risks. This has been made all the more necessary with the introduction of massive deep-draughted ships which have little room to manœuvre in channels and which, as a result of their great mass, cannot be brought to a standstill in an emergency in less than quarter of an hour.[8] Ships between 200,000 and 300,000 dwt, powered by steam turbines, will cover about 1·5 miles with engines going full speed astern before they can be stopped from an ahead speed of 16·5 knots and, as they lose speed in attempting to stop, they also lose the ability to manœuvre. The cost of collision is nowadays so high, £10 million to £20 million for the total loss of a giant ship, and the pollution, fire and explosion

World shipping routes

hazards from oil, liquid gas and chemicals are so great, that ships will accept compulsory routeing along certain lanes even if it means the lengthening of time on passage by several hours and loss of customary freedom of the seas.

There have, of course, for many years been such recommended tracks separating east- and west-bound vessels crossing the north Atlantic. Weather routeing of ships on this run will make such tracks less meaningful. But in actual practice there is relatively little need for anti-collision tracks in the open sea. Where compulsory tracks are very necessary is at the approaches to land and in channels and straits. As a measure of the seriousness of the collision problem Stratton shows that the annual loss per 1,000 tankers (over 500 dwt) rose from 2–3 in 1959–64 to 3–8 in 1965–8; most of the collisions can now involve very big ships.[9]

There are other divisions of the high seas which affect the movement of shipping. For safety purposes the oceans are divided, by international agreement, into seasonal load-line zones. These demarcate the oceans in such a way that a vessel passing from one seasonal zone to another must not be loaded below the corresponding seasonal marks shown on the vessel's hull, deeper loading being allowed in summer compared with winter, or in tropical compared with temperate zones. This may affect choice of route. For example, a decision may have to be made between adopting a great circle route across the northern Pacific at winter marks, that is following the shortest distance with less cargo, or a rhumb-line route further south at summer marks, that is a longer distance with more cargo. The load-line zones also affect the ports of call. On a voyage which begins in a winter zone and ends in a summer zone the owner or master may, after considering freight rates and bunker prices, decide to take maximum cargo and minimum bunkers in order to bring the ship to the required winter marks. In this case he would then call at the first suitable port in the summer zone to load the rest of his bunkers. Ports near the junction of seasonal zones, such as Las Palmas, Cape Town and Colombo, have undoubtedly gained in the numbers of ships calling as a result of such voyage patterns.[10]

It might be noted that since the 1966 revision of the load-line rules the Cape of Good Hope now lies in the permanent summer zone; prior to this ships rounding the Cape of Good Hope during the southern winter were required to be at winter marks. The 1966 rules, which allow loading to summer marks throughout the year, represent a substantial gain in cargo for large vessels on this route and will undoubtedly add to the comparative advantages of the Cape route *vis-à-vis* the transoceanic canals.

1. Burger, W., and Evans, S. H., 'Heavy Weather at Sea. The Efficiency of Weather Routeing in reducing damage', *Journal of the Institute of Navigation* **24**, 3 (July 1971), pp. 284–90.

2. Svendsen, Arnljot Stromme, *The Northern Sea Route and Its Future Importance to International Sea Transport and Shipping* (Bergen 1963).

3. Winstanley, J. D., 'Surveying Deep-Draught Routes', *Journal of the Institute of Navigation* **23**, 4 (October 1970), pp. 411–16.

4. Panama Canal Co., *Annual Report* (1967–8) and *Panama Canal Review 55th Anniversary* (1969).

5. *Nord-Ostsee Kanal, Quarterly Reports*. Federal Ministry of Transport (Kiel 1960–9).

6. Baxter, R. R., *The Law of International Waterways* (Cambridge, Mass. 1924), p. 96.

7. McDougal, M. S., and Burke, W. T., *The Public Order of the Oceans* (Yale 1962). Also de Blij, H., *Systematic Political Geography* (New York 1968), chs. 10 and 11; and Alexander, L. M., 'Geography and the Law of the Sea', *Economic Geography* (March 1968).

8. Tani, H., 'On the stopping distances of giant vessels', *Journal of the Institute of Navigation* **23**, 2 (April 1970) pp. 196–211.

9. Stratton, A., and Silver, W. E., 'Operational Research and Cost Benefit Analysis on Navigation with Particular Reference to Marine Accidents', *Journal of the Institute of Navigation* **23**, 3 (July 1970) pp. 325–40. Also Paffett, J. A. H., 'Technology and Safe Navigation', *The Institution of Engineers and Shipbuilders in Scotland*, Paper 1362 (November 1971).

10. *The Merchant Shipping (Safety) Load Line Rules* (HMSO, London 1968).

5

TRENDS IN MODERN SHIPPING

As was observed earlier, under the *laissez-faire* conditions of the nineteenth century Britain stood supreme in shipping and at the turn of the century still possessed almost half the aggregate world tonnage of merchant vessels. The overseas coal trade continued to provide considerable employment for British ships at this time, for out of the world total of 250 million tons of goods transported by sea in international trade about 80 million tons was coal from British ports.

Many nations expanded their fleets during the early years of the twentieth century and by 1919 Britain's share of world tonnage was reduced to 34% while that of the United States stood at 20%. Both coal and cotton began to show a marked decline in British trade, and growth in world trade as a whole was slowed. The world merchant fleet continued to expand over this period and its efficiency was continuously improved; this meant overtonnaging and difficult times for shipping, although subsidies helped sustain the fleets of the United States, Italy, Germany and some other countries, but not those of Britain, Greece and the Scandinavian nations.

Since the interwar period there has never been the same dominance of one nation in world shipping. The world fleet has grown from 60 million g.r.t. in 1939, to 100 million in 1959, 211·7 million in 1969 and 247·2 million in July 1971, while the tendency has been towards a greater spread of national ownership. Fast rates of merchant fleet growth have taken place in Japan, in some of the socialist countries, and under flags of convenience. The Japanese fleet increased from two million g.r.t. in 1950 to 30·5 million in 1971 and this rapid build-up means that most of the ships in the fleet are less than ten years old. The fleet of the USSR is also new and has risen in the world shipping league from twenty-first place in 1950 to sixth place in 1971. The US fleet shows only a slow rate of growth, but much of the tonnage registered under the Liberian flag is owned by United States companies. India is the only nation amongst the developing countries to possess a significant merchant navy; although in Africa, south-east Asia, South America and in China there have been relatively rapid developments of small fleets. Table 1 shows the gross tonnage of the principal national fleets in July 1971.

Table 1
WORLD SHIPPING BY FLAG, JULY 1971

	Gross (000 g.r.t.)		Gross (000 g.r.t.)
Liberia	38,552	Denmark	3,520
Japan	30,509	India	2,478
UK	27,335	Canada	2,366
Norway	21,720	Poland	1,760
*USA	16,266	Brazil	1,731
USSR	16,194	Yugoslavia	1,543
Greece	13,066	Cyprus	1,498
Germany (W.)	8,679	Finland	1,471
Italy	8,139	China (Taiwan)	1,322
France	7,011	Argentina	1,312
Panama	6,262	Belgium	1,183
Holland	5,269	Australia	1,105
Sweden	4,978	China (Republic)	1,022
Spain	3,934	Germany (E.)	1,016

*Including USA Reserve Fleet (5 m g.r.t.).

Source: *Lloyd's Register of Shipping Tables* (1971).

Demand for shipping

The demand for merchant shipping is derived from the volume of international trade generated between countries which can be directly, or indirectly, linked by sea. The variations in amount and direction of this trade are in turn dependent on a complexity of short-term factors, particularly changes in political conditions and climatic variability, so that demand for ships appears difficult to predict. On the other hand, there are several long-term trends, such as world population increases, growth in the gross national product of various countries, and rising standards of living, at least for the majority of people in the advanced manufacturing countries, which have contributed so far to an almost continuous expansion of world trade and hence to increases in the demand for merchant ships. The growth in world seaborne trade has been of the following order, in million metric tons:

	Dry cargo	Tanker cargo	Total[1]
1950	300	225	525
1960	540	540	1,080
1969	1,189	1,091	2,280

The demand for sea transport is determined not only by the quantity of cargo to be moved in international trade but also by the distances over which it has to be moved. Any change in the components, tons or miles, alters the demand for shipping. It was, for example, not so much the decline of coal exports from Britain in prewar years which contributed to the depression of the British shipping industry, but specifically the decline in the long-distance coal trades to Canada and South America. Conversely, an increase in ton-miles explains the greatly augmented demand for merchant ships during the first Suez Canal closure, for although the quantities of cargo offered in the world remained virtually the same immediately before and after closure, the mileage requirements by the alternative route around the Cape of Good Hope increased substantially, and so therefore did the number of ships required.

There has in fact been a tendency in recent years for world distances to increase in the haulage of raw materials by sea. This is related to the opening of new sources of mineral ores in Brazil, Africa and Australia, and especially to the expanded industrial capacity of Japan based on imports. The increases in ton-miles in world trade of iron ore and coal is made clear by Table 2.

Table 2

INCREASES IN TONS AND TON-MILES OF SEABORNE BULK COMMODITIES BETWEEN 1960 AND 1969

(mill. tons; thou.-mill. ton-miles)

		1960	1969	% increase
Iron ore	tons	101	214	113
Iron ore	ton-miles	264	919	248
Coal	tons	46	83	80
Coal	ton-miles	145	405	179

Source: Fearnley and Egers Chartering Co. Ltd (1970).

The tonnage of international seaborne trade will undoubtedly continue to grow in the immediate future, probably at an annual rate of 6 to 8%. The emphasis will obviously shift between commodities as standards of living change, as agricultural policies alter, and as new materials develop. The patterns of trade involving any particular commodity may also alter considerably with shifts in world supply and demand. Rising incomes in the developed countries tend, for example, to generate demand for more manufactured goods, much of which, with only marginal differences in quality and price, flow along main liner routes. In 1969 almost 55% of world exports by value moved between developed countries. Parts of the developing world may, by contrast, show a slower rise in the import of manufactured consumer goods partly due to tariff barriers to support domestic production, but also as a result of the increased price of manufactured goods *vis-à-vis* the price of raw material exports. The developing country may, however, go through an expanded phase of high imports of development goods in the form of machinery and bulky equipment. Increased demand for greater quantities of certain raw materials such as ores and timbers, in the manufacturing countries may compensate some developing countries, in purchasing ability, for rising prices of manufactured goods and hence retain their import volumes; but this rise in raw material demand is also likely to stimulate the search by the developed nations for cheaper raw materials and synthetics and thus bring changes in volume and areal shifts in patterns of trade.

The present imbalances in world seaborne trade, which are only partly accounted for by differences in resource endowments between countries, will thus be perpetuated for some time to come by economic trends. They are reflected in the different growth rates in import and export patterns of developed and developing groups of countries; taking cargo tonnages in 1958 as 100, for example, the growth trends to 1968 were as follows:

	Developed	Developing[2]
Growth in imports	217	167
Growth in exports	187	219

One can appreciate from these figures that many ships tend to leave the developed countries light in cargo loads. This is actually more pronounced than it appears since imports and exports are frequently carried by different types of ships. These imbalances in trade, and the tendency to move raw materials over greater distances, have given rise to attempts to minimise ballast voyages by designing ships capable of carrying foodstuffs, raw materials, liquid cargoes, cars, and other large items of manufactured goods, on alternative legs of voyages. The offering in this way of low cost carriage facilities for return cargoes also generates new trade, so that shipping to some extent can stimulate demand for its own services by these innovations.

Functions of shipping

The main function of the world fleet of merchant ships is to close the physical gap of marine space between production and consumption in the world economy. There is a vast amount of movement on the world's seas; on any one day there may be 10,000 to 15,000 ships crossing the oceans. This movement generally passes unobserved. Only at points of convergence, as in the Strait of Dover through which 800 vessels pass in a day, can the function of shipping as an important integrating factor in the world be appreciated; although when shipping is disrupted by military activities or strikes the significance of these services for food supply and industry is made more apparent.

The vital function of shipping in meeting the transport needs of a complex world economy can be appreciated even more readily from the fact that three-quarters of international trade by weight, and over 65% by value, moves by sea; and that in terms of freight rates shipping is still the least expensive method of transporting large quantities of goods over long distances. About half the commodities moving in world trade are tanker cargoes of crude oil, another quarter of the tonnage comprises iron ore, coal, grain, fertilisers, bauxite, timber and sugar, in that order of tonnage; these are all homogeneous cargoes transported by bulk carriers and tramp ships; the balance is made up of a vast array of manufactured goods shipped mainly by conventional cargo liners and unitised vessels.

Sea transport also functions in the movement of people. This too is basically an intermediate service dependent for its existence on numerous socio-economic activities. Passenger ships have engaged in the carriage of migrants from regions of food and employment shortages to regions of surplus land and a demand for labour; or in transporting passengers to places where they could pursue business or recreational activities. The first function has declined with the reduction in overseas migrations, and the second has been eroded by the faster linking facilities offered by air transport. The decline is most marked on the north Atlantic, the busiest of transoceanic passenger routes, where, as will be seen in Chapter 6, it has been both relative and absolute.

Where sea passenger transport provides a final service, in meeting and stimulating the demand for holiday and educational cruises, there are opportunities for new ocean passenger tonnage. Sea passenger services may also continue to expand with provision of roll-on/roll-off ferries, and other specialised types of vessels, on the short-sea routes where the speed of air transport has no decisive advantages.

Fleet structure

The changing emphasis in the functions of shipping are naturally reflected in the structure of the world fleet of ocean-going ships. This can be appreciated from Table 3 which shows the recent growth in oil tankers and bulk carriers and the comparative decline in other types of vessels. In 1970 oil tankers constituted 39% of the total world fleet, liner and tramp vessels 32%, ore and other bulk carriers 18%, combined carriers 4%, passenger ships 2%, container vessels 1% and liquid gas and chemical carriers 1%.

Table 3
WORLD FLEET 1964–72*

	Tankers		Combined carriers		Bulk carriers		Others	
	no.	mill. dwt	no.	mill. dwt	no.	mill. dwt	no.	mill. dwt
1/1/64	2,656	69·2	77	2·4	843	17·1	11,905	85·8
1/1/68	2,918	103·0	153	7·7	1,498	38·7	12,395	87·9
1/1/72	3,219	168·2	258	20·2	2,327	68·7	12,988	91·7

Source: *Fearnley and Egers Review* (1971), p. 9.

* Tankers, combined carriers and bulk carriers comprise vessels over 10,000 dwt, others comprise all seagoing cargo vessels over 1,000 g.r.t.

A real decline has taken place in passenger shipping, while conventional tramp ships and liner tonnages have been static. Tramp ships are being superseded in many trades by special bulk carriers, and liners replaced by container and other unitised vessels. There is a decided trend to specialisation in shipping with the building of chemical carriers, cement carriers, woodchip carriers, car transporting vessels and numerous others; in 1970 about 160 different types of ships were employed in seaborne trade. The new bulk vessels, such as giant ore carriers, are engaged in the transport of homogeneous materials between ports fitted with specialised handling equipment to provide rapid turnround, and the introduction of container ships is simply an application of the principle of bulking to general cargoes. In this case by packing, and unpacking, heterogeneous manufactured goods to and from a container while the ship is at sea, then loading and unloading the container at special terminals

Trends in modern shipping 75

rapid turnround times commensurate with those in the bulk trades can be achieved in the carriage of general cargo.

It is these trends in ship size and specialisations, with their matching facilities on shore, which have brought increased efficiency to modern shipping, and have raised the amount of cargo per ton of shipping space delivered per annum. They have also brought accelerated rates of ship obsolescence and the world fleet has, as a result, tended to become younger as replacement has proceeded. In 1955 18% of the fleet comprised vessels of over twenty-five years of age, by early 1965 only 7% of the world fleet was of this age and over. The new vessels have higher speeds at sea, better equipment for port working, and are less liable to lengthy periods of repairs; ship productivity is thus continually being raised.

Ownership and growth trends

Despite greatly improved efficiency in most of its sectors the world fleet as a whole has continued to grow in tonnage by about 10% per annum. This growth has been unequal between countries. As was seen in Table 1 (p. 70), Liberia has taken the lead in growth, but this particular tonnage group is owned largely by American and Greek companies. These owners have registered their ships under PANHOLIB (Panama, Honduras, Liberia) flags of convenience and have been able to speed the rate of ship acquisition through capital accumulated from untaxed profits. In the case of American nationals they have obtained in addition the advantages of lower crew costs compared with the high wages ruling on American flag vessels. The complexity in ownership and operation of vessels sailing under flags of convenience has been succinctly described as: an owner from country A, with money borrowed at advantageous rates in country B, operates vessels under flag C, from a management office in country D. The vessels can be chartered to owners in country E and manned by a crew from countries F, G and H.[3] About 15% of world merchant tonnage operates in this way, and this presents formidable competition to strictly national shipping companies subject to higher taxes and more stringent laws regarding manning and safety requirements.

Several governments in developed and developing countries are active in promoting the expansion of their national shipping by various means. They provide support by legislative edict which requires home shippers to use a certain proportion of national flag tonnage. National fleets are also supported by subsidies; either under a so-called infant industry policy (until the industry can stand on its own feet in competition with others), or as in the USA, to enable ships to compete with lower-wage operators. For this latter purpose the

American government pays subsidies of up to 50% of operating costs to certain shipping companies engaged on routes considered essential for the United States. The American government can also direct at least 50% of foreign-aid cargoes to US ships; these various measures keep about 80% of American tramp ships operating.[4]

Trade agreements between countries often involve stipulations on the use of national ships, and in some of the Latin American countries national ships are assured of cargoes by law. There appears, likewise, to be a high degree of cargo reservation in the COMECON group of countries. However, the Rochdale Report on shipping considers that, in 1966, only about 3% of the non-communist world's seaborne trade would have been diverted from normal commercial channels by flag discrimination. The extent to which Soviet ships operate under government subsidy is not clear. This fleet is now on world-wide services; in particular growth of trade has been very rapid to the Middle East, Africa, and south-east Asia. More than two-thirds of the ships have been built in the last decade, and between 1971 and 1975 the Soviet merchant marine will annually receive one million tons of shipping. Soviet authorities maintain that the fleet operates strictly on commercial lines.[5]

Most governments do in fact give support to their shipping. The British government has helped by depreciation allowances in respect of taxes, but Britain and some of the other maritime countries are opposed to direct operating subsidies, restrictive clauses in bilateral trade agreements, or discriminatory practices such as preferences in the berthing of vessels, or differentials in port charges. All support measures, direct or otherwise, encourage a general expansion of the world fleet.

Shipbuilding trends

Another force working in the direction of increasing the world tonnage is the soft loans offered by some countries with high shipbuilding capacity to secure orders for their industry. Shipowners may thus be induced to embark on building programmes not on the basis of what appears to be the demand trends in international trade but because of easy credit. Governments of most shipbuilding nations have, in fact, taken measures to persuade their own nationals to build at home and to induce foreign shipowners to place orders with their yards. Loans of 80% of the price of a vessel, repayable at $5\frac{1}{2}$% interest over ten years, have been made by Britain, and more generous terms have been offered by some other nations, with even longer repayment periods.

In addition to credit facilities shipowners are influenced by speed of delivery when they order ships. Some countries are at a disadvantage in this respect due to inadequate shipyards and labour problems. Many British shipyards, for example, are located on riverside sites in areas of nineteenth-century industrial and housing developments. There have been difficulties in altering their layout to facilitate computer-controlled linear flows of materials as ships are constructed. Such a change in the morphology of a shipyard implies also changes in patterns of work, for under modern conditions the shipbuilding industry takes the form of an automated assembly plant, but even where a new layout has been partly achieved the industry often retains many labour-intensive, multi-craft structures and relationships. These have led to demarcation disputes as workers have become concerned with possible redundancies as a result of the introduction of new work methods and equipment; such problems have plagued the shipbuilding industry in Britain and parts of Europe.

The Japanese, by contrast, have had several advantages in shipbuilding and have, as Table 4 shows, become the leading shipbuilding nation in the world. This may be ascribed to the tremendous demand for ships by Japanese nationals; but Japanese yards have also succeeded in attracting one-third of all the shipbuilding orders for western European countries during the last decade. They have done so partly through government assistance to shipbuilding but also as a result of very efficient new shipyards established on greenfield sites since 1950.

Table 4

WORLD SHIPBUILDING 1952–69

(mill. g.r.t. per annum)

	UK	West Germany	Sweden	Japan	World
1952	1·3	0·5	0·5	0·6	3·4
1962	1·1	1·0	0·8	2·2	8·4
1968	0·9	1·4	1·1	8·6	16·9
1969	1·1	1·6	1·3	9·4	19·5

Source: *Lloyd's Shipbuilding Statistics* (1969).

Most of the new shipyards in Japan have been laid out on reclaimed land, and labour has been drawn from rural communities in the areas. The yards are all part of bigger industrial groups, such as Mitsubishi which owns steelworks, overseas ore deposits, and ships, as well as five shipyards. This type of ownership structure allows each yard in a group to specialise in a certain class of vessel. The building processes are highly automated and the industry is big enough to exert great pressure on subcontractors to ensure prompt delivery of components at the correct stages in construction programmes. In 1971 the Japanese had plans for six new shipyards and aimed at building 75% of world tonnage.

The trend in Britain and Europe is towards a modernised shipbuilding industry. Some yards have been established at greenfield sites; others involve giant building docks, as at Belfast where ships of over 500,000 dwt could, if necessary, be constructed. These developments are likely to lead to over-provision in world shipbuilding capacity; this will be accompanied by severe competition, possibly more government support for national shipbuilders—certainly encouragement for shipowners to build ahead of demand, and, for developing countries in particular, the ordering of vessels for new national fleets.

The importance of national shipping

Despite what might be considered as artificial inducements to increase tonnage the rates of expansion in national fleets do indicate something of the importance of shipping for the countries concerned. As a transport facility it is clearly significant in any economy for most goods embody some element of direct or indirect sea transport costs. These include freights in the import of foodstuffs and raw materials, which can add anything from 10 to 50% to delivered prices, or freights on processed and manufactured goods to the prices of which up to 10% may be added.

It is understandable from these very general estimates why countries wish to reduce the costs of sea transport whenever possible. One method of so doing is to employ the most economic carriers offered on the world transport market. But it is also clear that there are several countries which feel that a shipping industry is so important for them that they must purchase and retain their own tonnage, even if the ships cannot pay their way in a competitive world situation and have as a result to be subsidised.

At first sight this appears irrational. Shipping is, after all, a service which can be bought and sold on a national or international scale. As with other economic activities some countries have acquired comparative advantages in producing shipping services, and unless there are radical changes in the factor requirements for efficient operation they may be expected to continue to specialise in shipping. Other nations may purchase such services and concentrate on activities in which they, in turn, have comparative cost advantages. This, in theory, is the principle of specialisation and the geographical division of labour, on the basis of which much of the world economy has advanced.

There are obviously, in this respect, nations with a history of maritime enterprise, a legacy of ports, ships, shipbuilding, ship-repairing and marine engineering facilities and skills. Such countries also possess other long-term acquired endowments such as maritime education and research institutions, and marine insurance companies, as well as brokers, experienced seamen, managers and shipowners. A shipping industry, in effect, is made up of a complex of interacting activities. These are difficult and expensive to acquire, but without them efficiency in shipping is equally difficult to achieve. But the fact is that even highly efficient merchant fleets do not earn as much for their services as they did fifty years ago. Indeed the purchasing power of Norway's shipping services in terms of Norway's imports has declined by as much as half over that period. This would have been a serious economic trend were it not offset by the increased productivity of Norwegian shipping. Tresselt has remarked on the irony of the situation whereby developing countries are trying to enter an enterprise which by many criteria a modern industrial country should be trying to get out of. And Sturmey says that if Britain did not have a merchant fleet it would be unwise to acquire one.[6]

Investment in shipping has nevertheless certain economic and non-economic attractions for some countries which seem independent of accounting profitability. The industry may, for example, earn scarce foreign currency by transporting goods between overseas countries, and it can conserve foreign exchange by substituting freight payments to national flag ships instead of to foreign ships.

The balance of payments attraction of shipping is undoubtedly a strong motive for embarking on shipping activities. There are also other reasons of a less quantifiable nature. These include the prestige factors in operating national flag ships, also the view that foreign shipping companies may be overcharging and, in any event, that political and economic independence requires national shipping. Or, in the case of oil-producing countries, it is yet another way of entering the downstream activities of marketing. Then there are defence arguments: not only does the existence of a national fleet provide a reserve of trained seamen, and vessels which can be utilised for the transport of troops and materials, but without national ships there is always the possibility of serious shortages of tonnage if foreign maritime nations are involved in wars. Finally, there are the views that ownership of vessels can promote overseas exports, or that it can assist the economic integration of trading bloc areas, and that shipping is a source of national employment. Shipping thus appears as a valuable asset which, seemingly, is independent of natural resources. There is some validity in many of these arguments and they are referred to again in Chapter 9. The prime importance of shipping for most countries, however, rests on balance of payments contributions. National shipping can sometimes help solve balance of payments deficits even if it remains an enterprise which requires financial support. Investment in shipping does, of course, divert capital from other enterprises which may serve this purpose equally well, or even better, by promoting exports or by import substitution. Also, in adopting a shipping solution to balance of payments difficulties, there may be the equally difficult problem of purchasing the ships and their equipment overseas which can represent a drain on foreign currency reserves.

The real value of shipping to a country is complex, as many be appreciated from the example of Britain summarised in Table 5.

Table 5
UK SHIPPING CONTRIBUTION TO BALANCE OF PAYMENTS 1969

Credits	£m
Freight on exports	196
Freight on cross trades	501
Time charter receipts	61
Passenger receipts	64
Disbursements by overseas ships in UK	137
Total credits	959
Debits	
Time charter payments to overseas shipowners	210
Disbursements by British ships abroad	371
Freight on imports by foreign ships	318
Passenger fares on foreign ships	15
Total debits	914
Net gain to balance of payments	£45 million

Source: *United Kingdom Balance of Payments 1969* (HMSO 1970), p. 68.

It will be noted that freight earned by UK ships for the carriage of imports to Britain is not included in the credits. The reason being that for balance of payments purposes these are treated as domestic transactions. This is one of the reasons why the contribution of £45 million to the balance of payments, valuable as it was, did not reflect the true value of the British merchant fleet to the economy. Had the fleet not existed then foreign exchange earnings on cross trades would have been lost and almost the whole of the UK overseas trading activities would need have been to conducted by foreign shipping. The UK would have saved financially from avoidance of overseas disbursements incurred by her merchant fleet, and would have gained from the added expenditure of more foreign vessels in UK ports, but a balance of payments deficit of over £400 million might have resulted had all cargoes been carried by foreign flag vessels. As Britain exports about £8,000 million in goods and services per annum, and suffers not infrequently from adverse balance of payments, a decrease of this magnitude in the shipping account would be serious, and would also mean reductions in the industrial and commercial activities which support the shipping industry.

In the long run the actual effects of not possessing a merchant fleet would depend on the alternative uses to which the resources of capital and labour were put.[7] The apparent attractions are such, however, that many nations see solutions to their overseas currency problems in this way. Developing countries, for example, generate between them 41% of the world's seaborne trade (64% of loadings and 18% of unloadings, 1969) but they collect only 5% of the gross earnings of world shipping. Some of these countries naturally look to earnings from the carriage of their own exports as a means of obtaining and conserving overseas funds. The Latin American countries in 1961, for instance, recorded a deficit of $527 million in the shipping accounts of their balance of payments. This was a compulsive force in the decision to expand and protect, by cargo preferences, the fleets of the Latin American Free Trade Area.[8]

Shipping organisation and markets

By according cargo preferences to their national fleets governments are, in effect, attempting to create their own closed markets for shipping. But in spite of these methods, and the many bilateral trade agreements which contain restrictive clauses, ocean shipping is still an international business subject to vagaries arising from world supply and demand conditions; the organisation of shipping markets and management reflects this fact. Internationalism is most pronounced in the open market where world shipowners are usually seeking cargoes for their ships and many shippers throughout the world are looking for suitable vessels on which to ship their goods. The London Baltic Exchange and the New York Exchange are the centres at which this information is gathered, and where the available ships and cargoes are matched by brokers bargaining on behalf of the shipowners and the charterers.

Both parties in these exchanges have a sound knowledge of supply and demand conditions for ships and cargoes. Should there be a shortage of tonnage in one region then high rates are offered to attract ships away from other trades. As this demand is met the rates fall in the initial attracting area and may rise elsewhere. But a real equilibrium of price, representing a balance in supply and demand of ships, is seldom if ever met with on a world scale. Owners with ships in a location where there are few cargoes may have to accept very low rates or proceed on long ballast hauls; others can secure exceptionally high rates by having their vessels free and at the right place at the right time, particularly during periods of sudden increases in the demand for tonnage arising from political or meteorological factors which affect commodity supply or demand.

The dry cargo ships which are placed on the open market are tramps available at short notice for the carriage of all kinds of commodities to and from any part in the world. Their services may be sold under voyage charter agreements, whereby the vessel undertakes to deliver a certain tonnage of cargo to specified places at a fixed payment per ton. The earnings in this case are quite closely related to the amount of cargo carried and to the distance steamed. Once the cargo is delivered the vessel is free to commence another charter which will have been fixed to take whatever advantages there may be in the ship's geographical location. If the voyage is to terminate in an area with few return cargoes the shipowner will attempt to obtain rates or terms in the charter to compensate for this.

Another type of fixture is the time charter, under which a ship is hired for a period of several months or years to trade between specific areas at a rate per deadweight ton per month, or at a fixed rate per day. It is the ship's capacity which is being hired in this case irrespective of the amount of cargo hauled or the nature of the voyages, although the shipowner may stipulate extra charges in respect of dangerous or corrosive cargoes, or in relation to certain routes or ports.

Both the voyage and the time charter rates are determined by supply and demand for commodities, and by business expectations. Fixtures are made in the competitive open market mainly in respect of tramps, small bulk carriers, and oil tankers; and increasingly for the big combined carriers that can shift between oil, ore, and bulk cargoes as market conditions change. Because of the many variables involved in world trade there can be striking variations in the rates offered on the voyage charter market even over a few days. The time charter rates show more stability, but the two markets interact and are subject to decisions based on the owner's market perception and the degree to which he is risk-prone or risk-averse. Should world trade show signs of slowing down, for example, many shipowners try to obtain long-term fixtures rather than risk voyage chartering in a falling market. This tends to lower the rates being offered on the time charter markets; and of course opposite conditions prevail when trade appears buoyant. Some owners are prepared to gamble and operate their ships on the 'spot' voyage charter market most of the time, others divide their fleet between the two.

In general, voyage chartering is adopted in trades with seasonal peaks, in trades which tend to be erratic in supply and demand, and in cases where more tonnage is called for to meet marginal increases in the requirements of liner services. Time chartering is very common in trades which require the delivery of large quantities of bulk commodities at regular intervals to meet the more certain demands of heavy industry. Three other methods are used to ensure this regularity. First, the contract of affreightment, whereby a shipowner, or ship management company, undertakes to supply the shipper with tonnage, rather than any particular, ship, on a regular long-term basis. Second, the bareboat or demise charter where the charterer (an oil company for example) takes over the vessel and runs it as the owner; and third, where ships are actually owned by very large vertically integrated companies. This latter form of industrial shipping is found mainly as subsidiary branches to the oil and steel industries, but it is also increasingly met with in the chemical, sugar, and other large-scale industrial enterprises dependent on a steady flow of raw materials from overseas sources.

In contrast to the erratic nature of the tramp market the liner market is relatively stable, as freight rates for cargo liners are less directly affected by the supply and demand for commodities. Liners once established in a trade are expected to operate scheduled services irrespective of the amount of cargo offered, and they are characterised by receiving small shipments from a great number of shippers; their commercial activities are thus in the nature of common carrier contracts.

The tariff charges in the liner market for a wide variety of goods are actually agreed on between the various liner companies operating services to the same area under conference, and sometimes in consultation with shippers' organisations; but seldom do short-term changes in demand for the commodities carried affect these rates. On the other hand, the amount and type of cargo which liners receive can be affected by the level of rates they quote. Liner cargoes can also be affected by the supply of tramp ships available for charter to big shippers or groups of shippers, and sometimes by other transport modes, in particular by air freight. These various markets and ships are considered again. They are all influenced in some ways by world commodity supply and demand. In turn the volume and direction of flow of goods, especially primary products, can be affected by the freight rate ruling at any one time.

Individual shipowners cannot influence world commodity supply or demand to any extent. Nor have they any real control over the deployment of the world supply of ships, or therefore on freight rates, other than when operating in closely knit conferences.

All that the owner can do to ensure a good average level of earnings, over poor and favourable markets, is continually to improve his efficiency and reduce operating costs to a minimum. Attention to costs is clearly of vital importance in what is an internationally competitive industry. Any tendency for one element of ship costs to rise must be offset either by trying to secure higher freight rates, which is often difficult, or by offsetting the rise by improved efficiency or technological advance. Another alternative is to secure a subsidy; but because subsidies are often a drain on national resources, and inhibitors of efficiency, they are frequently opposed as solutions to rising costs in ship operating.

Costs in shipping

The costs in sea transport are frequently divided into three broad categories: capital costs, running costs, and voyage costs. Both capital and running costs are fixed; that is they are costs which are inevitable once a voyage has commenced irrespective of the areas of operation or whether the ship is carrying a full cargo or is proceeding in ballast. By contrast voyage costs cover variable expenses which are determined by the nature of the voyage, that is they may vary according to the cargo loaded, the ports of call, the route followed, or the weather conditions encountered during the voyage. Under fixed capital costs come depreciation and the interest payments on the initial capital involved in building the ship; under fixed running costs are included such items as wages, stores, insurance, surveys, repairs, indemnity and protection club payments, and management. The variable costs are composed mainly of cargo handling costs, port charges, canal dues and fuel.

The fixed and variable cost characteristics of marine transport have an important bearing on the nature of competition within the various markets. In the liner trade, for example, because a high proportion of costs are fixed there is little added to total costs if, when the ship is partially loaded, additional cargo is accepted at rates which are a little above the variable cost of handling the cargo. This is one of the reasons why liner companies insist on freight-rate agreements between ships trading regularly in an area, for under free-market conditions rates could be driven to low levels and the ships could not for long maintain the scheduled services on which many shippers depend.

In the tramp market the fact that a relatively high proportion of costs are fixed has the effect of causing owners to delay laying-up their ships when rates are low. It will cost them money if they continue to operate on a low freight market but it will also cost them money if they lay the ship up, for the depreciation and some other items of

Trends in modern shipping 85

fixed costs must continue to be met. This explains why owners with new vessels have less incentive to withdraw ships from service, for they cannot save in capital costs by so doing, and newness is a feature of the world fleet. The result of these characteristics is that surplus ships are kept operating at a loss for much longer than would otherwise be expected, and this can render a depression in shipping long lasting.

Shipowners may also be reluctant to lay-up their ships because of obligations to crews, or because they would lose credibility with shippers or financiers, or simply because of lost prestige. Mainly however, the decision is made on strictly economic grounds. When, for example, the total operating costs minus the likely freight earnings are greater than the cost of taking the ship out of service, maintaining it, and recommissioning it, then a ship may be considered for laying-up; shipowners will, in other words, run the ships at freight earnings below operating costs by as much as the cost of laying them up.

There is a close interaction between market demand, costs, and freight rates. Svendsen illustrates this as in Fig. 2. Here OA represents the quantity of tonnage in a given market, the curve TT' represents a series of total sailing costs per deadweight ton for different ships on line OA. The curve SS' represents a series of freight rates on the laying-up point per deadweight ton for the same

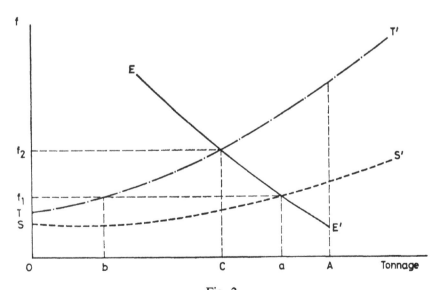

Fig. 2
Freight rates and laying-up of ships. Source: Svendsen, A. S., *Factors determining the laying-up of ships* (Bergen 1957).

ships. The curve EE' is the demand curve for tonnage. If the level of freight rates lies at f_1 only Oa is employed while aA is laid-up. Of the ships which lie between A and O, only those represented by Ob operate with a profit. If the available tonnage in the market were Oc the freight rate would be fixed at f_2 and almost all ships would sail at a profit. Fig. 2 thus shows the effect of overtonnage on freight rates, and unless ships are withdrawn, or the world trading situation improves, it can become chronic.

Clearly, the shipowners with lowest costs and most efficient vessels can survive for longest when markets are poor. In attempting to reduce costs shipowners in countries like Britain have concentrated on certain elements. Crew costs, for example, have risen appreciably and have been the subject of control exercises. For a typical British cargo ship crew costs came to £18,000 in 1954, £25,000 in 1957, £29,500 in 1960 and £40,000 in 1962.[9] They have now been offset, to some extent, by increased size of ship which has improved the manpower/tonnage ratio. Labour costs on a 25,000-ton ship, for example, were £2·50 per deadweight ton in 1969 compared with £0·30 per ton on a 250,000 tonner in 1970. Substitution of mechanical equipment for labour, and the more efficient deployment of manpower, has allowed crews to be reduced to the minimum laid down for safety.

Repairs and dry-docking costs have also tended to increase. Reductions have been made in these by the adoption of scientific protection devices against corrosion, and by the use of new constructional materials such as high-tensile steel. It might be noted that the latter also reduces the overall weight of material in the ship and thus allows more cargo to be carried; this is particularly important for tankers and iron-ore ships which normally load to their marks.

Of all the above measures the economies of scale are the most important in leading to reductions in costs. This may be appreciated from an examination of Table 6 which shows the estimated capital costs and running costs of three British bulk carriers of different sizes on charter at 1968–9 prices. It does not show cargo-handling and fuel costs which were borne by the charterer.

The cost variations shown in Table 6 are worth further consideration as the spreading of costs over larger transport units in this way is the motive behind many developments in modern shipping. It may be seen that while the capital costs of building ships increase with size of ships, and so therefore do depreciation and interest payments, they do so at a diminishing rate per deadweight ton. Clearly on these figures, a 150,000-ton ship could be built for £5·5 million, whereas it would cost £9 million for six 25,000-tonners in order to provide the same aggregate cargo capacity. This economy arises from a saving in labour and management employed in building the big vessel (some tasks are not very sensitive to size) and, more significantly, from a relative reduction in the amount of materials used. With larger holds, or tanks, there are proportionately fewer subdivisions; while the space for the crew, and for engine-room and navigational equipment, remains about the same as on the smaller vessel. Furthermore, there are few additions to auxiliary machinery with increased tonnage, up to about 300,000 tons.

Table 6

COMPARISON OF COSTS OF BULK CARRIERS ON CHARTER 1968–9

000 dwt	25	100	150
Capital cost (£m)	1·5	4·5	5·5
Crew costs (£000)	59·5	78·5	79·4
Provisions stores and lub oil	21·4	32·8	33·7
Insurance	18·7	102·8	125·7
Indemnity payments	2·2	14·7	18·0
Repairs	9·7	18·0	20·0
Management	7·0	10·0	10·0
Dry docking	5·0	8·0	9·0
Spares	3·3	7·5	8·5
Hires	3·2	3·7	3·7
Sundries	2·0	2·0	2·0
All other	4·0	10·0	10·0
	136	288	320

Source: J. and J. Denholm Ltd (1968).

Neither do crew numbers increase with size of ship; they may in fact decline with the automation which is possible in larger units. This is reflected in the reduction of men employed in the British merchant navy from 150,000 in 1957 to 97,648 in 1970, while the total tonnage of the fleet has increased from 19·8 mill./g.r.t. to 25·8 mill./g.r.t. over the same period. Management costs and staff may also be reduced by the same process, for it is clearly a simpler task to manage a small fleet of big vessels than the converse. When it comes to insurance and indemnity payments, however, there are substantial cost increases with size, as underwriters have become increasingly concerned with the magnitude of claims for ships, cargo, and damages, when a giant vessel is lost or is involved in a collision.

Elements of variable costs, such as port expenses, are not always so firmly in control of the operator, but in the case of fuel there are once again economies to be derived from size of ship. For vessels operating at a constant speed, for example, there is a reduction in tons of diesel oil used per deadweight ton in the following proportions: 35,000 dwt, 55 tons per day; 50,000 dwt, 63 tpd; 75,000 dwt, 80 tpd; 100,000 dwt, 100 tpd.[10] This saving continues, at a diminishing rate, until ships of over 300,000 tons are reached, when two propellers may have to be fitted, with a consequent higher consumption of fuel as well as increased capital expenditure.

It will be appreciated why, as a result of the economies in scale, a 200,000-ton tanker can deliver cargoes at less than half the cost per ton of a 30,000-ton vessel over a distance of say 8,000 to 10,000 miles. The most marked trend in modern shipping is therefore towards increased size. If the physical conditions of ports are suitable, if large quantities of cargo can be received by consignees, and if the ship can obtain rapid despatch, then the tendency is to continue to build on as large a scale as possible for any particular trade.

The geographical consequences of achieving cost reductions through economies of scale have been many, and these will be referred to more fully in the chapters which follow. But we might note that a large highly capital-intensive ship is built primarily for moving cargoes from one port to another over long distances. The more trips it can make the greater are the profits for the shipowner when demand is high, and the more competitive he becomes when the market is poor. As port turnround improves, and as more of these ships come on the market, the average freight rates tend to drop. As all giant ships can operate at lower freight-rate levels these rates become the norm for all vessels. In this way the world economy benefits in the long run from the search for economies in shipping; the enterprising shipowner benefits by being first in the field but in the long term he merely retains his position, hence the continued stimulus to improve.[11]

The impact of the giant ship is felt not only on the commodity market via freight rates but also at the ports. If such a vessel is delayed in port its high depreciation charges continue while its potential for earning is under-utilised. It is then, in effect, acting as an expensive warehouse and not doing the main job for which it was built. Port equipment and layout must therefore change in phase with growth in the size of ships. Shipowners, for the reasons stated above, will undoubtedly continue to set the pace in these developments, and industry and ports will adjust in order to retain or acquire competitive advantages. Trends in ship size and type are consequently amongst the most potent of the forces acting towards geographic change on a world scale. The principal types of ships are discussed in subsequent chapters in this context.

1. UNCTAD TD/B/C.4/66. *Review of Maritime Transport 1969* (UN, New York, 1969), p. 3.
2. ibid., pp. 4–5.
3. Buzek, F. J., and Wepster, A., 'The Master in Changing Times', *Journal of the Institute of Navigation* 24 (1971), pp. 35–42.
4. Fair, M. L., and Reese, H. E., *Merchant Marine Policy* (Cornell U.P., Ithaca 1963), and US Department of Commerce, *Essential US Foreign Trade Routes* (Washington 1960).
5. Bakayer, V., 'The Soviet Merchant Marine', *Shipbuilding and Shipping World* (6 September 1968), pp. 294–6, and Harbron, J. D., *Communist Ships and Shipping* (London 1962).
6. Tresselt, Dag, *The Controversy over the Division of Labour in International Seaborne Transport* (Inst. Shipping Research, Bergen 1969), p. 5, and Sturmey, S. G., in *Shipping the next 100 years* (J. and J. Denholm Ltd, Glasgow January 1966).
7. Rochdale Report, *Committee of Inquiry into Shipping* (HMSO, Cmnd. 4337, London 1970) pp. 342–60, and Goss, R. O., *Studies in Maritime Economics* (Cambridge 1968), pp. 46–60.
8. Brown, R. T., *Transport and the Economic Integration of South America* (Brookings Institute 1966), ch. 6.
9. J. and J. Denholm Ltd, information 1968.
10. Colley, E. G. S., *The Economics of Large Tankers* (BP Tanker Co. Ltd, 1965), p. 9.
11. Bolton, F. B., 'The Economics of the Bulk Carrier', *Journal of the Honourable Com any of Master Mariners* (December 1969), pp. 16–17.

6

CONVENTIONAL DRY CARGO SHIPPING

A conventional ship is the type of dry cargo vessel that appeared with the improvements to the steam reciprocating engine in the late nineteenth century. It is equipped with its own cargo handling gear and is of a size which enables it to enter a great number of the world's ports. Many of the ports were, in fact, designed, or redesigned, to accommodate conventional vessels, which are divided basically into tramps and liners. Until recently conventional ships were the backbone of the merchant fleets of the world. Although since their early days these dry cargo vessels have undergone improvements in speed and construction they are not so radically different now from what they were fifty years ago. This chapter discusses conventional ships and indicates the forces at work which are leading to a diminishing role for them in world economic geography.

Tramp shipping

Modern tramp ships can be defined by their function only, for size and appearance mean very little. The tramp is a dry cargo vessel operating a world-wide service under charter to the nationals of any country, other than those with whom economic relations are prohibited by government edict. It tends to concentrate on the carriage of relatively low-value homogeneous commodities usually on behalf of one consignee.

Tramp ships first appeared on the world trading scene about the mid-nineteenth century. They were introduced by British shipowners and their development was stimulated by several interrelated factors. The repeal of the British corn laws in 1846 initiated an era of free-trading, the opening of the Suez Canal in 1869 confirmed the advantages of steam over sail, the rapid growth of population and industry in Britain and Europe gave concomitant demands for foodstuffs and raw materials, the construction of railways opened the interior of North and South America as sources of grain, and the new telegraph communications from 1860 onwards made possible the transmission of information on commodity demand and ship positions necessary for chartering. Britain by the mid-nineteenth century was already building steamships on the Clyde and Tyne, and the coalmines of these areas supplied cheap bunkers and coal exports, while the Empire, as well as providing a market, offered a unique chain of bunkering ports and naval stations around the world.

Most of the British tramps of the late nineteenth century were vessels of 1,000 to 2,000 tons deadweight owned by numerous small companies in north-east England, Glasgow, Cardiff and London. They carried the heavy and bulky commodities required by the industrial nations, consequently the main direction of cargo flow was towards western Europe. But British owners, as well as being first in the field with these ships, had a singular advantage, for they could almost always obtain outward loads of coal. This enabled them to quote lower rates on some cross-trades and on the homeward legs of voyages, especially with grain. Coal out and grain home, with intermediate trips on cross-trades, was standard practice for British ships. The rival Scandinavian maritime nations without coal resources found this type of competition particularly difficult and they attempted to compete on long voyages with low-cost sailing ships.

Coal continued to dominate world fuel consumption throughout the interwar years and British tramps dominated world bulk trades. But the change from coal to oil in world industry and transport which has taken place since the interwar period has reduced the competitive advantages once derived by Britain from cheap bunkers and outward loads. The Scandinavians, in particular, were quick to take advantage of oil fuel. Diesel ships require a lower tonnage of fuel, hence there is more cargo space, and the engine-room wage-costs of diesel vessels are about a quarter of those on a steamship. The fact that British owners continued with steam vessels until the Second World War contributed to some of the subsequent difficulties of the British shipping industry.[1]

The decline of British coal exports from 100 million tons per annum hauled before the First World War to less than 4 million tons per annum in recent years has meant that most tramp ships must nowadays proceed outwards from Britain in ballast, and much of their earnings are made on competitive cross-trades. Comparative advantages in operating relatively small tramps lie no longer with cheap bunkers but with countries having low crew costs.

In order to spread crew costs and achieve other economies of scale the tendency has been to increase the size of tramp ships. Indeed, it is very difficult to distinguish between tramps and bulk carriers hired on the open market. For some statistical purposes tramp shipping of the present day has been taken to include any dry cargo vessel of over 4,000 dwt regularly engaged in deep-sea tramp trading, except those adapted to a single commodity.[2] This definition excludes single-purpose ore carriers and other specialised vessels, but includes a wide range of general-purpose bulk carriers. Tramp ships so defined totalled 54·7 million dwt in 1967, or about 20% of the world fleet. The principal tramp fleets are shown in Table 7.

Table 7
PRINCIPAL TRAMP FLEETS 1967

	No. of ships	mill. dwt	% World
Greece	806	10·4	19·1
Norway	408	8·5	15·4
Britain	433	6·6	12·0
Japan	262	4·3	7·9
Italy	213	3·5	6·4
USA	191	2·6	4·8

Source: *Westinform shipping report* **280** (1968).

The trends already noted in world shipping to increased ship size and more specialisation has meant that the *raison d'être* of the free-ranging tramp has been eroded, especially by bulk carriers operating regular services on long-term contracts. The latter ships have tended to capture most of the increases in the trade in bulk commodities. The general-purpose tramp is still predominant in the grain trade, although in some years oil tankers have taken one-quarter of the American grain exports. The independent tramp has also continued to provide ready tonnage to meet sudden increases and seasonal expansions in both the special bulk carrier and the liner markets. Tramps, in other words, continue to provide an essential flexible element in world shipping which is capable of relatively rapid transfer between trades and able to serve a great range of ports. Vessels constructed to a standard design in order to replace flexible wartime Liberty ships, such as the British SD 14 with a tonnage of around 14,200 dwt, a draught of 8·68 m (28½ ft), and a speed of 14 knots, are still in demand for these purposes.

The pattern of tramping
In operating tramp ships the shipowner attempts to arrange his

Conventional dry cargo shipping

charters so that vessels move around the world minimising ballast-voyages and following the seasonal shifts of supply and demand.[3] He seeks to terminate each charter at a place from which a full cargo can be secured at high rates. This is seldom in fact possible, for owing to directional imbalances in world trade return cargoes may be difficult to obtain. As a result of this between 30 and 40% of the time at sea of a modern tramp ship is spent in ballast.

Some of the seasonality which once guided the pattern of tramping has been reduced in recent years but, as can be seen from Fig. 3,

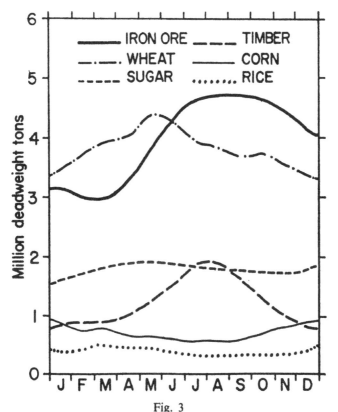

Fig. 3
Seasonal fluctuations in bulk cargoes (averages 1948–58). Source: Fearnley and Egers and Norström (1961), p. 122.

there were still quite distinctive seasonal peaks in the demand for shipping over the period 1950–7. In the carriage of wheat there was a spring high and winter low, the iron-ore trade tended to rise from a winter low to a summer plateau, while the timber trade showed a sharp rise in spring and a rapid decline after midsummer. Coal for domestic purposes still occasionally experiences a winter rise if weather conditions are particularly severe in Europe, but in general seaborne coal is for industrial purposes and industry requires regularity of coal deliveries.

It will be appreciated from the areal and seasonal demand for tonnage in the leading bulk trades that tramp ships could usually expect favourable market conditions during the northern summer. But now the main seasonal highs based particularly on spring and summer harvests cannot be relied on. This is especially true of the wheat and corn trades in which tramp ships are still the most active participants. In 1966 FAO estimated that 837 million tons of grain was produced in the world and of this 76 million tons moved by sea. But the following year world exports were reduced by 13%, and in 1968 by a further 9%. Good domestic harvests in India and Pakistan, the recovery of Russian grain, and the large stocks built up in other countries resulted in reduced tramp-ship activities.

Some of the advantages of arranging ship movements to be at forward loading positions when regional shipments of grain start have consequently also been reduced. There is no longer the same coincidence between the North American summer harvest and peak autumn shipping activities in North American ports. Storage facilities have regulated the flow of grain more in accordance with market requirements. The rush to get summer-harvested grain out of the Great Lakes before the winter closure is also less urgent. At one time tramps could obtain very high rates by risking ice-capture in the Great Lake ports while loading the last of the season's shipments, but now downstream elevators in the St Lawrence hold stocks of grain which they release to meet market demands throughout a longer navigation season.

It is still possible to identify an element of harvest-induced movements in the grain trade of the southern hemisphere. Grain shipments from Argentina, for example, tend to reach a peak in January, offering a counterbalance to any lulls in northern cargoes; tramps can thus move southwards in winter. This movement of tonnage can also be achieved in relation to Australian wheat; but here modern elevator facilities once again allow the overseas market to predominate over the harvest season in determining the demand for shipping. Storage has thus regularised the flow of grain throughout the world, and has eliminated the situation whereby the existence of large newly harvested grain crops seeking outlets meant that growers had to bear a considerable proportion of the high freight rates and at the same time received reduced prices for their crop. But this reduction in the seasonal amplitude of freight rates has also lessened the opportunities for tramp ships to balance bad with good periods of earnings.[4]

Tramp ships still find summer charters in the flush of timber exports from high-latitude coniferous forest zones of Europe and North America. The ports at the northern end of the Baltic are the first to freeze and the last to thaw so that the timber exporting season is short and busy in this area. It is partly a reflection of the lower value of sawn timber, compared wih paper and pulp, that it is stockpiled for the arrival of the tramps in spring instead of being sent south by land transport to ice-free ports.

The seasonal rhythm of the timber trade by sea reflects in fact diverse aspects of the human geography of high-latitude countries.[5] With the winter lull in northern agriculture there is a transfer of labour to timber cutting, the frozen terrain in turn allows a reduction in land transport costs by enabling the timber to be hauled to rivers and loading places for spring export; and in the importing countries, where the building industry is at a low ebb during the winter, there is a corresponding seasonal pattern of timber utilisation.

The demand for shipping generated by the seasonal activities in timber means tramp ships can obtain good freights during the spring and summer timber surges, although ports may be congested and ship delays result. Because of port delays, the restricted access to shallow loading ports in the Gulf of Bothnia, the use of the Kiel Canal, and the relatively short distances involved, Baltic timber ships are generally of the 1,000 to 3,000 dwt class. This also allows these timber carriers to call at many small ports in the importing countries, thus avoiding congested areas. By contrast several of the timber vessels from the more distant White Sea ports are around 6,000 dwt and call at fewer ports, while those employed throughout the year on the British Columbia–UK timber trade are even bigger and more specialised, reaching 30,000 dwt. They deliver packaged timber to three ports only in Britain for onward shipment by road and rail to national hinterlands.

Tramp earnings

The seasonal and other variations in cargo flows in the tramp market naturally affect freight rates. When several commodities are competing for carriage, high shipping rates are quoted to attract ships away from one trade, or area, towards another. As noted, however, the storage and release of grain on a more regular basis tends now to reduce the amplitude of freight rate fluctuations in this particular trade, and as grain is the staple tramp cargo this affects the tramp market as a whole. On the other hand, unexpected political and weather conditions continue to give rise to exceptional demands for tramp ships from year to year. Examples of this can be observed in the inflated rates current during the Korean war, and during the exceptionally high demand for US coal in Europe in the winters of 1954 and 1957. Fig. 4 shows the rate fluctuations in the important grain trades between 1963 and 1968.

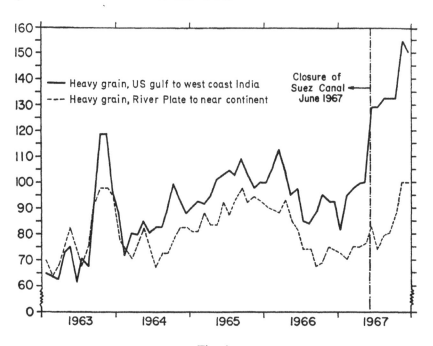

Fig. 4
Average monthly freight rates in (shillings) grain trade, per long ton (1963–8). Source: UNCTAD TD/B/c.4/38 (1969), p. 85.

It is virtually impossible for the tramp owner to predict these unusually poor or favourable markets, but he responds to trends. For example, when freight rates are falling a tramp owner may attempt to obtain long-term time charters before the rates drop too far. But a temporary fall in freight rates towards the level of uneconomic operating does not normally result in any significant increases in cargo tonnages, or in the generation of many new cargoes which could redress earnings. It is, however, also fortunate for the tramp owner that a temporary rise in freight rates to levels which are highly profitable does not result in a serious loss of cargo. These characteristics are due mainly to the inelastic nature of the demand for commodities such as grain and ores carried by tramp ships; they are so vital for industry and for the support of urban communities that their supply must continue to be assured. In the case of copra, and some natural fibres, demand is more elastic. An increase in freight rates, raising the c.i.f. price of such commodities, could result in a reduction in demand by manufacturers as cheaper natural and synthetic substitutes are brought into use. In practice this might only be so if the increased charges were passed on to the buyer but in many instances they are passed back to the producer and represent a lower f.o.b. return to him for his products. It will be appreciated from this alone that the attempts to achieve economies in shipping, which make lower freight rates acceptable to shipowners, are of particular importance to primary producing countries.

Because of this relatively inelastic character of demand for certain primary commodities, tramp shipowners attempt to make the best of periods of high freight rates. They may do this by increasing the round voyage speed of their ships to a maximum and may choose to ballast back to the areas where high freight rates are ruling rather than spend time loading and unloading lower-rated cargoes, although this may eventually raise rates on the latter. High rates will also bring most laid-up ships on to the market. Once all seaworthy vessels are employed, and if the demand continues to increase, the freight rates will continue to rise even more steeply until shipbuilding programmes, initiated by the boom, produce additional tonnage. On the other hand, when there is a surplus of shipping and freight rates fall, an owner may have to choose between accepting the losses from operating at low rates or the losses from laying-up his ship. Frequently old ships will be the first to leave the market. These tend to have higher operating costs and lower capital costs compared with new vessels, and since it is mainly operating costs that are saved by laying ships up the more capital-intensive vessels are often kept running. Much depends, however, on size and type of vessel, on rates of wages, and on the estimates by individual shipowners of future prospects. An owner may also continue to operate at a loss in order to retain the goodwill of a regular charterer, and always in the hope that conditions will improve.[6]

Tramp prospects

As we have seen tramp shipowners depend over the long term on the steep increases in freight rates which result from shipping shortages. With the present speed of shipbuilding such boom conditions last now for shorter periods. Also, the very variability in supply of ships and the related fluctuations in rates have induced companies relying on regular deliveries of raw materials to secure their own supply of shipping, either by direct purchase or by long-term charter. By so doing they have reduced the opportunities for tramp ships. This is most apparent in the iron-ore trades where vessels on long-term charters are now the main carriers. In turn, long-term employment has allowed specialisation of ship types and accompanying economies of scale, thus rendering the general-purpose tramp even less competitive.

There is little doubt that the tramp market has suffered in this way, but with the continued growth in world trade the loss has been so far only relative. Tramp ships hired on the open market have retained about 15% of world trade[7] and they continue to predominate in grain shipments, although tankers and some bulk carriers have challenged the general-purpose tramp even here. Nevertheless, the 'handy-size' ship of around 10–18,000 dwt, with its own cargo gear and moderate draught, will continue to be required for trading to the many hundreds of ports in the world with restricted entry, and to areas that generate cargo at infrequent intervals and in variable quantities. But little significant expansion is likely in this type of conventional tramp shipping. It must be concluded that more than ever the tramp market is full of uncertainties and requires a good deal of individual judgment on the part of the owners. The only safe method of operating is to secure long-term charters, but this means reduced opportunities of very high profits during sudden periods of boom in world trade.

Types of conventional liners

Throughout the period of industrial growth in the nineteenth century tramp ships proved invaluable in meeting the needs of industry for raw material imports. But tramp shipping services were inadequate for the export requirements of manufacturing industries. Many manufacturers required frequency of deliveries and regularity of contact with a wide area of markets, thus allowing relatively small quantities of high value goods to be delivered on predetermined dates and so minimising stockholding and maximising capital turnover.

Cargo liners were developed to meet these needs. The various liner companies established scheduled services to certain ports, and as a result of regularity of trading they began to adapt their ships to the special characteristics of the trades. A combination of higher speeds at sea, more elaborate cargo gear to reduce port time, separate compartments for the carriage of special products, an increased number of intermediate decks, and the provision of refrigerated space came to typify liners. Although most of them were built mainly for the carriage of general cargoes outwards from the manufacturing countries, a number were designed principally with the inward trades in view—such as the refrigerated vessels on the meat and dairy produce trades and on the banana runs. Other liners returned to Europe with consignments of rubber, tea, coffee, tobacco, cotton, jute, palm oil, kernels, and similar relatively high-value primary products, as well as some grain and quality timber. These ships in meeting the requirements of a particular region made available suitable cargo gear, including heavy lift derricks, and special stowage facilities for a range of commodities, including bulk liquids, fibres, timbers, metals, chemicals, explosives, fruit, fragile goods, banknotes, mail and frozen products in a variety of packages of various shapes and sizes. Most liners were fitted with humidity control systems or special ventilation to protect cargo from damage on passing through different climatic zones.

In their early days liners provided a combined service with cargo, passengers and mail. A gradual separation of passenger services from cargo activities took place over the years, although until about 1965 most cargo liners retained passenger accommodation for up to twelve people now very few do so; but passenger ships, especially those on long voyages, continue to provide a little cargo space. The dual functions of cargo and passenger carriers proved uneconomic for two reasons: first, cargo work could delay the ship from sailing at a fixed time, thus defeating the requirements of a passenger carrier for all possible regularity of departures and arrivals; second, the dual-purpose ship had to engage a large complement of catering staff which was underemployed during the long periods of loading and discharging of cargo, when passengers were not normally on board. Liner shipping thus divided into two functional types: specialised passenger and mail-carrying ships running rigorously to schedule between a limited number of ports, and cargo vessels with somewhat more flexible schedules covering a greater range of ports.

The separation of functions in liner shipping was soon reflected in size differences. By virtue of the ease and speed with which passengers could be embarked and disembarked a single-purpose passenger ship could be rapidly turned round in port. The only constraints on its size were technical problems relating to stability, the limits of the canals, and the number of passengers available on any run. Passenger ships were thus very early in the twentieth century built to sizes above 30,000 g.r.t., and they continued until recently to be the biggest ships afloat. The cargo liner could not grow in this way. Because of the many diverse items carried the loading and unloading proceeded slowly. A very large cargo liner would have resulted in an extremely protracted period in port. In these circumstances some of the first items loaded might be in the ship for months, with a consequent deterioration of condition and a loss of working capital for the owners of the goods. The size of a conventional cargo liner was, therefore, dictated by the character and requirements of the market for the goods, and by cargo handling equipment and port conditions. The lack of growth opportunities due to these constraints now place conventional cargo liners amongst the smallest of ocean-going vessels.

Operating cargo liners

If we accept the definition of the conventional cargo liner as a multi-deck dry cargo ship operating a regular service, then they constituted about 12% of world's shipping in 1969, or about 26 million tons gross, of which the UK operated 6·2 million tons and was the principal liner-owning nation.

The modern cargo liners of the conventional type, that is distinct from container ships, are between 10,000 and 13,000 dwt. Because most of their cargoes are of relatively light weight the ships are built with as large a cubic capacity as possible, commensurate with hulls designed for speed. However, there is considerably broken stowage on a general cargo liner due to the varied sizes and shapes of packages, and on refrigerated ships the depth of insulation also lowers the space available for cargo. Further streamlining of modern cargo liners to attain higher speeds at sea, thus compensating for time lost in port, has moreover reduced cargo-carrying capacity. These features of the liner, the high standard of equipment, and their powerful engines giving speeds of 20 to 27 knots render them an expensive class of vessel.

Cargo liners provide a service by calling regularly at specified ports irrespective of the quantity of cargo available. Their schedules are, of course, related over the long term to cargo expectancy. Usually, the amount and frequency of shipping on any trade is determined by that leg of the voyage which offers the highest cargo potential, for there are almost always imbalances in cargo flow even in the trade between developed countries. In the trade to a developing country seasonality of primary production for export is frequently met with and adds to the complications of providing adequate tonnage; unless, for example, the seasonal surge is in that leg of the voyage which normally has a low load factor, then the liner company will have to charter additional shipping in order to meet obligations to shippers.

The problem of scheduling liner services through low and high periods of cargo availability adds to a related problem in liner operations, that of the rates to charge shippers for services. There are so many common fixed costs in operating a liner, ranging from depreciation to management, that it is impossible to allocate them between the commodities carried in a way which might reflect costs of carriage. It is possible to allocate some of the variable costs, such as cargo handling, in this way but not with any degree of accuracy.

Conventional dry cargo shipping

The high fixed costs also mean that when cargo flows are reduced this is not compensated by commensurate savings in the costs of operating the ship.

Since it is not possible to spread common costs, and it would be cumbersome to alter freight rates according to the amounts of cargo carried on each voyage, some basis other than cost-related charges has had to be adopted in liner operations. One general solution has been to adopt a value-related system of charging. High-value goods carry high freight rates and low-value goods low freight rates with, in addition, surcharges on goods of a dangerous nature, awkward size or shape, or destined for particularly difficult ports.

This practice of charging what the traffic will bear requires freight-rate agreements between liner companies. This is necessary because the marginal costs of carrying additional items of cargo are so low that were there no such agreements there would be considerable temptation for a ship to accept cargo at quite low rates on occasions. This would undoubtedly favour shippers over the short term but in the long term would damage their interests. A freight war could mean some ships would be driven out of the trade, possibly those with expensive facilities for the cargo; it would certainly mean less reliable services since at low rates ships would be willing to call only when sufficient loads were available, although this tendency might have the effect of raising rates for small shipments very considerably and the small ports and shippers would suffer on both counts. The fluctuations in rates and the variability of services which would occur as a result of lack of agreements between liners would mean shippers having to carry greater stocks of goods, and some export-oriented manufacturers being unable to sequence their output in relation to market demand and transport flow. It may also follow from this that ultimately the company which could withstand low rates for the longest period would capture the trade, and under monopoly conditions would raise rates to a higher level than those ruling under the conference system.

A number of shippers would for these reasons appear to favour a system in which reliability ranks high and they are prepared to pay for this. Some shippers have, however, reservations about certain aspects of the so-called conference system which the liner companies have evolved to ensure both reliable services and viable operations. As the conference agreements are amongst the principal guiding elements in liner operations, and their workings have considerable geographical implications, we must look at them in detail.

Conferences[8]

Agreements on the scheduling of services, the allocation of ports of call between conference members, and the fixing of freight rates, have characterised liner services almost since their inception. On the UK–India trade in 1875 agreements were already being made between shipping companies for the joint regulation of freight rates. This system spread rapidly to every major liner trading area in the world. The liner conferences established rules governing freight rates and made contractual agreements with shippers in order to ensure that they used the regular liners which were often specially designed for the trades. Inducements were also given aimed at retaining the loyalty of shippers by a system of deferred rebates, that is on each occasion a merchant shipped goods he received a rebate in respect of a previous shipment, and further measures were adopted to prevent non-conference ships interloping. Sometimes these involved drastic price cutting, or the threat of this, until the interloper was driven out after which normal rates were re-established.

The conference system ostensibly offers shippers regularity of services and stable freight rates as against spasmodic sailings and fluctuating rates. But conferences have met with the criticism of monopolistic tendencies and restrictive practices which maintain freight rates at unnecessarily high levels, and yet do not offer the best possible services. Also, certain conferences have been accused of influencing the economic growth of countries by their rates and other discriminatory policies. Leubuscher[9] instances a case in point from the west African trade. In the period between the wars an oil mill was established in west Africa for the processing of palm kernels; the west Africa conference refused to provide tank space on their ships for the oil exports or allow the chartering of a suitable vessel; as a result the enterprise failed. After the Second World War the palm-oil interests purchased their own vessels. It is also maintained that conferences have influenced economic development by favouring certain ports against others, either by frequency of service or by freight-rate discrimination, and the higher rates charged on manufactured goods compared with raw materials may have been a factor favouring market areas in developed countries for industrial location when an industry has had a free choice between locating at the raw material source or at the market.

Competition does, of course, exist within most conferences, but not in terms of freight rates. Vessels compete in speed of delivery, care of cargo, and reliability in such matters as meeting claims. As a result of this type of competition high standards of service are achieved. But this policy has also been open to criticism since it raises costs, and hence rates, and the shippers of some cargoes may well prefer lower freight rates to very high standards of service. Competition may also exist between different conferences, as, for example, when ships are carrying the same commodities to a country from different regions of the world. It may be possible for one conference to render imports from its area of trade relatively more attractive by reducing the freight rates, thus improving on cargo volume carried by its ships, but common membership of several conferences by shipping companies minimises this type of competition.

Most discussions of shipping conferences turn on the subject of level of rates. The conferences maintain that enlightened self-interest would ensure against overcharging since very high rates might price a good out of the market and hence damage trade; it might even attract powerful companies aimed at breaking into the conference, or, nowadays, would give opportunities for air-freight services to capture high-value commodities; and, even more likely, any rates which appear to shippers to be excessively high may induce the country to acquire its own, even small, fleet as a countervailing force to the conference. It is, in fact, often only when the latter course is adopted that a country can ascertain with any degree of certainty whether the ruling freight rates are necessary for the maintenance of services. Conferences may also experience competition from specialised vessels such as timber carriers which can lower freight rates further on some of the low-value goods, and from tramp ships which may be chartered by big shippers or combinations of shippers. There are also many items of cargo, such as bagged grains and fibres, for which liners have normally to compete with tramps and do so at market rates. Conferences are now increasingly under pressure from these various competitors. In the carriage of high-value commodities competition comes from air transport and for some of the low-value commodities from specialised shipping and chartered tramps. There is, however, still a vast, and increasing, range of manufactured goods, primary products, and fresh foodstuffs carried by cargo liners.

Cargo liners: problems and prospects

The main defects of the conventional cargo liner in the present age clearly stem from the combination of the high capital content of the ship and the labour-intensive methods of cargo handling between ship and shore. Because of the latter an expensive vessel may be held in port for more than half of its voyage, and of this in-port period about half can be idle time when no cargo is being worked.

The difficulties of reconciling the high operating costs of the ship with protracted periods of cargo handling have contributed to the increases in liner freight rates in recent years. Between 1950 and 1968 liner rates rose on average by 80%, while the rates for most bulk cargoes fell by over 20%. Liners lost some heavy bottom cargoes to bulk carriers and special ships, and the rise in liner rates made substitution of air freighting attractive to some shippers, particularly since air rates were reduced over this period by 2 to 3% per annum. In Britain over 13% of exports by value now go by air and this is likely to increase with the introduction of all-cargo jumbo-jet services with relatively low ton-mile operating costs. Even where liners retain a price advantage over air freight the speed of capital turnover made possible by the latter renders it a viable competitor for goods in which the transport cost component represents only a small percentage of the value. It will be recalled that under the conference system of administered freight rates these high-value commodities are precisely the goods that subsidise the carriage of lower-valued products on which liners charge low freight rates. The liners depend on the high-value items and many shippers of low-value goods depend on liners obtaining these. For these reasons any tendencies to cost-induced freight-rate increases, especially arising from protracted port times, coupled with the erosion of earning opportunities due to loss of cargo to air transport, have been matters of considerable concern for liner companies and shippers alike. They have led to a critical re-examination of the conference system and of conventional cargo liner transport technology as a whole.

In improving the conferences the trend has been to take shippers more into confidence by producing conference shipping accounts. In the South African conference this is required by law and the government of that country in return will take measures against non-conference lines dumping their services at low freight rates in the South African trade. In the New Zealand conference an annual review of freight rates takes place between the representatives of the New Zealand Produce Boards and the shipping companies after an independent accountant has examined the voyage returns for the previous year. In the USA conferences must be open to all ships, but they are exempt from anti-trust laws in as much as they are allowed to fix levels of freight rates but cannot discriminate in price between shippers, or offer deferred rebates.[10] There is thus a tacit recognition of the value of the conference system, and a degree of approval for it when consultation is seen to be practised, or where conference operations are tempered by government regulations in the user country. There are about 360 conferences in the world and both developed and developing countries have membership of these, as have the ships of the Soviet Union and some other socialist countries. It appears that until a better system is evolved, or until the methods of carrying general cargo are completely changed, the conference system will persist.

Conventional dry cargo shipping

The conventional cargo liner will not persist on every trade route. The contradiction between high speeds at sea, which have been obtained by considerable capital investment in the ships, and slow labour-intensive handling of cargo in the ships' holds and on the quays, has brought about a need for the application of more technology along the ship-shore interface. A solution has been found in containerisation. This has changed both ships and ports and has many other implications which will be discussed in some detail in Chapter 7. There are, however, routes on which conventional cargo liners may remain for some years, such as the trades between developing countries which have many ports of call, the efficient South African fruit trade, and possibly the New Zealand meat and dairy trades. Also, with the build-up of general cargo flow in the world, and the ageing of many conventional liners, there could conceivably be a shortage of ships of the conventional class in a few years' time, thus stimulating a renewed building programme for cargo liners, especially ships of less than 10,000 dwt.

Operating passenger liners
The bases for high-capacity regular passenger services were laid during the period of the great migrations from Europe, Britain and Ireland to North America between about 1815 and 1900. Thereafter the gravitational effect of the two population masses of Europe and America separated by 3,000 miles of Atlantic Ocean generated an enormous force for interaction. This, allied with the cultural ties between the descendants of immigrants and their former homelands, led to the north Atlantic crossing becoming the primary passenger route of the world.

To meet the demand for travel across the Atlantic several fast passenger liners of more than 28,000 g.r.t. were developed from about 1900 onwards. Speed was an essential ingredient in these services and cargo space was reduced to minimise delays in port. The awarding of mail contracts and subsidies reinforced the importance of speed, and the growth of outports with railhead connections emphasised that the minimising of travel time was of primary importance for the transatlantic traveller, even in a more leisurely period.

With its emphasis on speed and comfort the passenger liner has always been an expensive vessel to build and operate. Nowadays depreciation, crew costs, and fuel can total about 80 to 90% of total costs. By contrast, costs attributable to the actual carriage of passengers, as distinct from sailing without them, are probably in the region of a mere 10% of total costs. Low load factors on passenger vessels mean, therefore, a reduction in earnings with little compensating reduction in costs.

Load factors have been reduced in recent years due to the faster transit facilities offered by jet air services. This has meant a reduction in the number of passenger ships operating in the world. Between 1958 and 1969 the tonnage of world seagoing passenger shipping fell from 8,163,000 to 6,734,000 g.r.t. The British fleet experienced the most marked decline from 2,511,000 to 1,415,000 g.r.t., the French fleet fell from 655,000 to 357,000, the Netherlands from 682,000 to 313,000 and W. Germany from 219,000 to 98,000.[11] By contrast the passenger fleets of lower wage countries such as Italy, Greece and Portugal, as well as the USSR, registered an increase in tonnage.

The decline in sea travel which has reduced the need for passenger ships is most apparent on the north Atlantic where, as Table 8 indicates, it has been accompanied by a general growth in the

Table 8

PASSENGER TRANSPORT ACROSS NORTH ATLANTIC 1957–69

	Sea (000)	Air (000)	Air (% of total)
1957	1,036	968	48
1958	957	1,193	55
1959	880	1,367	61
1960	865	1,760	67
1961	782	1,919	71
1962	814	2,272	74
1963	788	2,422	75
1964	712	3,069	81
1965	649	3,611	85
1966	606	4,198	87
1967	506	4,987	91
1968	376	5,259	93
1969	335	5,996	95
1970	249	7,202	97

Source: OECD, *Maritime Transport* (1970), p. 125.

numbers of people travelling between Europe and North America. Another problem of the passenger ship trade is seasonality. On the north Atlantic route decline in total passenger demand sets in from September. The effect is marked on sea transport so that by mid-winter there are no transatlantic passenger ship sailings from Britain, and very few from other ports in Europe, although a few cargo liners offer berths to would-be sea travellers. The north Atlantic, never a very comfortable sea crossing, has thus lost almost all winter passenger ship sailings to air transport. Passenger services by sea have also been drastically reduced in the links between Europe, India, and the Far East. In this case the closure of the Suez Canal accelerated the process of decline and now few people travel to these areas by sea from Europe.

Air services have had a less dramatic impact on sea passenger traffic on the more favourable weather routes between Italy and South America, and to some extent also between UK/Europe and Australia. Assisted passages for emigrants with their household effects southbound from Britain and Europe to Australia, and young Australians northbound at relatively low fares, have helped preserve some of these services. But air competition has eroded specifically the first-class sector of sea passenger transport to Australia, and as a result many of the vessels on this trade are of the one-class type. The continuous competition from air transport appears now to be acquiring significance at all levels, for in 1969 more than 50% of travellers from Britain to Australia used air services compared with about 5 to 10% in 1964.

One of the problems of sea passenger services where they continue to operate is their labour-intensive character. The ratio of crew to passengers is about 1 to 2·5 on most liners. In order, therefore, to minimise the numbers employed labour-saving arrangements such as automated galleys and cafeteria services have been introduced; but countries with a plentiful supply of unskilled labour, which is easily recruited as catering staff, have obvious advantages in the passenger market. Converted troopships, aircraft carriers and ex-cargo ships employing relatively cheap labour (although tending to become more expensive) and often flying flags of convenience, have appeared on the emigrant and cruising trades in competition with traditional passenger ship companies. However, in 1968 Britain still owned 21% of world passenger shipping, while the USSR, Greece, the USA and France owned between 5 and 6% each of the world total. Most of the American vessels received government subsidies without which they would not have been able to operate. It should be noted that the number of United States passenger vessels engaged in cruising is far less than the available cruise business in the United States, a fact which has helped countries such as Italy and Britain in this market. The maritime unions in the USA are, however, greatly disturbed by the decline of passenger ships of all types and have been pressing strongly for even more government support in the restoration of American flag passenger services.[12]

There is now, as a result of air competition, a market shift of emphasis from scheduled passenger services to cruising amongst all passenger-ship operators. This change in function has altered once more the types of passenger vessels being built. The very large ship is unsuited to cruising since it cannot always berth at the many ports which attract the holiday passenger, nor can it expect several thousand cruise passengers throughout the year in a highly competitive market. These are some of the main factors which are reducing the size of passenger vessels from capacities of over 3,000 passengers to about 1,000; the days of the giant passenger ships are clearly coming to a close. The British *Queen Elizabeth 2*, of 65,863 g.r.t., launched in 1968, with accommodation for 2,000 passengers and 1,000 crew, may be the last of the big ships.

The cruise ships must of course compete with air transport for business. Air travel is thriving in the holiday markets since airways offer fast delivery to attractive areas and often arrange hotel accommodation. The cruise vessel may, by contrast, take several days to reach the desired climatic zone. To overcome this shipping companies are integrating with airways in arranging charter flights carrying passengers to ships cruising in tropical and Mediterranean areas; this allows a more rapid turnover of cruise passengers and hence cheaper cruises. The Pacific is also proving a popular cruising zone. Several of the ships that at one time made regular passenger runs to and from the Pacific area now carry emigrant and tourist passengers outwards from Europe simply to take up their position for six months' cruising in the Pacific before returning with passengers to Europe.

Cruising appears to be the future of ocean passenger vessels. It is observable that as average incomes increase in a country so also does the propensity to consume services, including those associated with leisure activities. The cruise ship may, for example, benefit from the current shift in desired holiday standards by a growing section of the British population away from Victorian seaside resorts to continental beaches; such people are potential cruise passengers. Similarly, on a world scale there are countries, such as Japan, where new socio-economic groups are reaching cruise income level; at the same time in the United States yet another change in patterns of work and leisure manifests itself in earlier retirements and this also encourages more cruise enterprises. Passenger vessels of the right size can clearly shift from transport functions to cruising, and they may order their activities so that they move between regions of the world on a seasonal basis.

1. Course, A. G., *The Deep Sea Tramp* (London 1960), and Sturmey, S. G., *British Shipping and World Competition* (London 1962).
2. Westinform, *World Tramp Fleet*, Shipping Report no. 280 (September 1968).
3. Gripaios, H., *Tramp Shipping* (London 1959).
4. Nosträm, G., 'Seasonal Variations in the Employment of Bulk Cargo

Tonnage', *Tijdschrift voor Economische en Sociale Geographie* (May 1961), p. 119, and Fleming, D. K., 'The Independent Transport Carrier in Ocean Tramp Trades', *Economic Geography* (July 1968), pp. 21–35.

5. Mead, W. R., and Smedo, H., *Winter in Finland* (London, 1967).

6. Thorburn, Thomas, *Supply and Demand of Water Transport* (Stockholm 1960). Also Svendsen, A. S., *Sea Transport and Shipping Economics* (Bergen 1958).

7. Cufley, C. F. H., *The Ideal Tramp for the 1970s* (British Sulphur Corporation 1966).

8. Marx, D., *International Shipping Cartels* (Princeton, New Jersey 1953). Also UNCTAD, *The Liner Conference System* (New York 1970), and McLachlan, D. L., 'The Price Policy of Liner Conferences', *Scottish Journal of Political Economy* (December 1963), pp. 322–35.

9. Leubuscher, C., *The West African Shipping Trade* (Leyden 1963).

10. Garter, N., *United States Shipping Policy* (New York 1956).

11. Rochdale Report, *Committee of Inquiry into Shipping,* (HMSO, Cmnd. 4337, London 1970), p. 86.

12. AFL–CIO Maritime Commiteet, *Pilot* (National Maritime Union of America, January 1971), p. 6.

7

OIL TANKERS

The substitution of oil for coal as a source of energy in maritime transport has already been referred to. This energy replacement occurred also in industry during the interwar years. As a result world use of mineral oil steadily increased from about 200 million tons in 1920 to 2,000 million tons in 1969. Of this latter figure more than 50% was moved by sea. The demand for mineral oil is expected to continue to grow at over 5% per annum for the foreseeable future.

As will be seen from Fig. 5, the demand has come predominantly from the industrial countries of Europe and Japan, while the supply, with the exception of North American oil, has been drawn mainly from non-industrial areas in the Middle East, the Caribbean, north and west Africa and south-east Asia. The wide separation of producing and consuming zones, and the continually expanding demand for oil, have led to a phenomenal increase in the transport requirements of the oil trade. In 1971 this was reflected in a sea transport performance of almost 7,000 thousand million ton-miles compared with 470 thousand million ton-miles in the movement of coal.[1]

In order to meet the oil transport needs of the industrial nations a constant growth has taken place in the world fleet of specialised oil carriers and in the size of the transport units involved. These trends are indicated by Table 9. From this it may be seen that the increase in total deadweight tonnage has been at a greater rate than the corresponding increase in ship units. Between 1961 and 1969 world tanker deadweight of ships above 10,000 tons rose by about 100% whereas the number of ships increased by less than 12%. This was due predominantly to growth in the capacity of tankers and improvements in their turnround times in port. As a result an

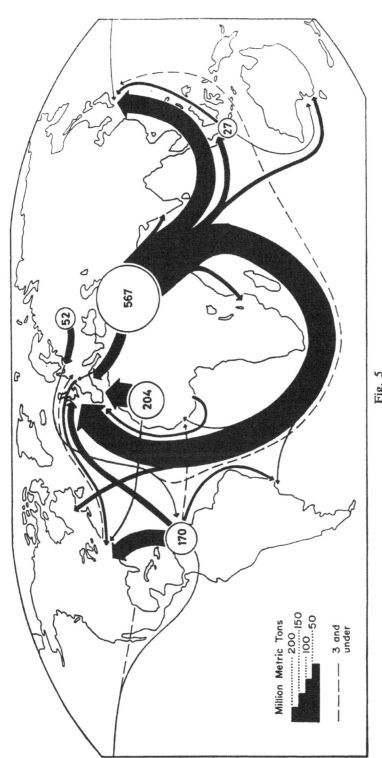

Fig. 5
Main inter-regional movements of oil by sea (1969). Source: *B.P. Statistical Review* (1970).

actual decline in the number of ships employed in the haulage of crude oil could set in.

The reasons why substantial increases in the size of oil tankers came only in recent years may best be appreciated from an historical perspective. During the interwar era there was a slow growth in the

Table 9

GROWTH IN WORLD TANKER FLEET 1900–69

	No. of ships		dwt
1900	109	above 2,000 dwt	531,000
1919	467	above 2,000 dwt	3,681,000
1939	1,571	above 2,000 dwt	16,600,000
1955	2,693	above 2,000 dwt	39,015,000
1961	2,671	above 10,000 dwt	60,616,000
1966	2,814	above 10,000 dwt	88,585,000
1969	2,991	above 10,000 dwt	121,016,000

Source: Petroleum Information Bureau (1969).

size of tankers from a maximum of 10,000 tons in 1919 to above 12,500 tons by 1939. The ships were designed for the carriage of a variety of refined products; but as they aged, and there was a consequent danger of contamination of products through oil leaking between tanks, many of them were transferred in their last years to the carriage of homogeneous crude oil. This latter activity was less important than the distribution of refined products.

In design most tankers from the 1930s onwards were subdivided longitudinally by two oil-tight bulkheads and transversely by several similar structures into a series of separate compartments. This ensured stability against the free-surface movement of liquid and also permitted the ships to carry several grades of oil at any one time. The relatively small size of tankers up to and including the Second World War allowed calls to be made over a range of the world's ports, and the tank subdivisions gave the ships flexibility in the parcels of cargo they could carry.[2] Flexibility was the keynote of these vessels; it was necessary at the time as three-quarters of the world's refining capacity was located in crude-oil-producing regions. The various products of these resource-based refineries, the gasoline, diesel, kerosene and fuel oils, were shipped widely throughout the world. The fact that few national markets could utilise the total product-mix derived from crude was another factor which tended to act against market-based refineries. India, and the other countries in the Far East and south-east Asia, required a high proportion of kerosene, Italy's main demand was for fuel oil, while a remote country like New Zealand, with few industries but a relatively high car-owning population, had an imbalance of requirements in the demand for gasoline.

The prewar ships were thus kept relatively small to perform product delivery functions and only a fraction of the tonnage was engaged in the carriage of crude oil at any one time. The latter served small home-based refineries in Europe and some installations at major bunkering ports such as Suez, Aruba, Curaçao and Singapore. The opportunity to separate crude from product carriers came only in the postwar period as a result of a marked shift in the location of refineries from the oil-producing to the oil-consuming regions of the world.

The change in refinery locations came about for four main reasons. First, there was the greatly increased demand in the industrial countries for all types of oil, for use in transport, electricity-generating stations, heavy industry, petro-chemicals and for thermal purposes. Second, refining techniques had advanced sufficiently to allow an output of grades of oil more closely related to local market requirements. Third, with technical advance there was no longer an excessive amount of unusable waste to be disposed of from a crude-oil cargo. And fourth, following the nationalisation of Anglo-Iranian refineries in Persia during 1951, there was the reinforcement of an already existing trend towards security of assets by building more home-based refineries.

The emphasis quite clearly shifted in the 1950s from locations aimed at flexibility in the distribution of refined oils to locations which offered flexibility in the import of crude. The new market-oriented refineries thus came to occupy positions in the industrial countries from which they could choose alternative sources of crude oil. Table 10 illustrates the accelerated growth in market-based refineries in Western Europe compared with the producing areas of the Middle East.

Table 10

GROWTH IN CRUDE OIL THROUGHPUTS IN WESTERN EUROPE AND THE MIDDLE EAST

(mill. tons)

	1939	1957	1959	1961	1963	1965	1967	1969
W. Europe	13	117	156	206	264	349	430	548
M. East	10	47	53	66	75	82	90	92

Source: *B.P. Statistical Review* (1969), p. 22.

Consequent on the shift in refinery locations the proportions of tankers engaged in crude oil and product haulage altered. About 75% of the tanker tonnage became crude-oil carriers and 25% was used for the distribution of refined oils, usually over short distances. Thus relieved of the need to haul varied products, which were limited by market requirements, it became possible to build large vessels with ten big cargo tanks, and subsequently with five tanks (15 compartments), simpler pipelines and pumping systems, and greater deadweight capacities. It was in the 1950s therefore that tankers began to increase appreciably in size. In 1951, for example, about 80% of tankers operating in the world were under 17,000 tons, by the end of 1956 half of the fleet was above this in tonnage and a few had reached 50,000 tons. It was simply the difficulties of quickly providing deeper terminal facilities in Europe, and the draught restriction of nearly 11 m (36 ft) in the Suez Canal (limiting loaded transits to vessels of below 45,000 tons), which held the size trends from accelerating at an even greater rate.

The closure of the Suez Canal in 1956 was the principal factor in removing one of the major constraints on shipowners to building very large vessels. That year saw a flood of orders for 100,000 dwt oil carriers for the purpose of hauling crude oil around the Cape of Good Hope. By 1963 there were vessels of 130,000 dwt in service and in 1966 ships of over 150,000 dwt were operating. Impetus was given over this period to the deepening of oil ports in Europe and to the provision of additional refineries at new deep-water locations. Table 11 summarises the increase in the proportion of giant tankers in the

Table 11

TOTAL WORLD TANKER FLEET 1938–69
(mill. dwt)

Size (000 dwt)	Approx. summer draught (metres)	1938	1958	1960	1964	1969
Under 25	9·75	16·6	38·7	36·8	31·3	28·9
25–45	9·75–11·89	—	15·2	21·6	25·4	26·7
45–65	10·67–12·48	—	1·2	4·3	16·1	22·0
65–85	11·89–1400	—	—	0·5	4·8	16·0
85–105	12·48–15·24	—	0·6	0·7	3·5	14·2
105–125	13·7–16·15	—	—	0·1	0·4	5·6
125–300	16.15–22·25	—	—	—	0·1	21·3
Total		16·6	55·7	64·0	81·6	134·7

Source: Institute of Petroleum (1970).

Oil tankers

world fleet from 1938 to 1969. In 1970 there were 131 tankers of more than 200,000 dwt in operation.[3]

The new very large crude-oil carriers have greater deadweight in relation to displacement, that is they can carry more cargo per ton of ship than the older and smaller vessels: a typical tanker of the 1950s, for example, would carry about 3 tons of cargo for every ton of ship, while the ratio on giant tankers is over 7 to 1. Yet on the latter ships loading and discharging times have been maintained at over 10% of the deadweight per hour; this has meant that unlike many dry cargo vessels growth in tanker cargo capacity has not imposed added port times. Almost all world tanker owners (including independent companies, government-owned fleets and oil companies) have now invested in very large crude-oil carriers.

Tanker owners

The ownership structure of the world fleet of tankers for 1969 is shown in Table 12. It may be seen from this that, next to Liberia

Table 12

OWNERSHIP OF WORLD TANKER FLEET AT END OF 1969
(2,000 dwt and over in mill. dwt)

Flag	Principal ownership			
	Oil company	*Private*	*Government*	*Total 1969*
Liberia	4·9	25·8	—	30·7
Norway	0·2	15·5	—	15·7
UK	13·6	4·7	0·3	18·6
Japan	2·2	11·5	—	13·7
USA	3·9	3·4	1·8	9·1
Panama	3·3	2·0	—	5·3
France	3·1	2·0	0·1	5·2
Greece	—	5·8	—	5·8
Other W. Europe	8·7	10·7	0·1	19·5
Other western hemisphere	2·5	0·3	0·3	3·1
USSR E. Europe and China	—	—	5·4	5·4
Other eastern hemisphere	0·6	2·2	0·1	2·9
Total	43·0	83·9	8·1	135·0

Source: *B.P. Statistical Review* (1969), p. 14.

(and the provisos that must accompany discussion of PANHOLIB 'ownership') Britain is the principal tanker-owning country in the world. This is a recent development and stems largely from the high proportions of tankers owned by British oil companies. Private tanker ownership by British shipping companies is still relatively small, a situation which may be an inheritance from the late entry into the tanker market by British shipowners as a result of their traditional bias for coal and steam. By contrast the Norwegians entered the tanker field in the days of the early diesel engine, and they thereby offset whatever disadvantages they had as a maritime nation without coal.[4] It was only in 1969 that the British fleet of oil tankers exceeded the Norwegians in tonnage. It should be recalled in this connection that the British government owns 40% of the shares of the British Petroleum Company which operates a major tanker fleet, also that tankers of the massive Shell fleet sailing under the flags of Britain and Holland belong to a company with 60% Dutch ownership and 40% British. Fine national comparisons of degrees of participation in the oil-tanker market are not therefore always very meaningful.

The international oil companies in Europe and America are in fact related in numerous ways by commercial agreements, and they periodically share shipping to minimise transport costs, especially by the reduction of cross haulage in world trading patterns. There is even some cooperation between capitalist and socialist countries in this way, as in the instance of the British Petroleum Company meeting Russian commitments to Japan by delivering crude oil from the Persian Gulf, and the Russians in turn meeting BP commitments by delivering oil from the Black Sea to the Mediterranean;[5] thus obviating the need for long hauls around the Cape of Good Hope by tankers travelling in opposite directions with cargoes of crude oil.

Taking the world oil-tanker market as a whole it is usual for oil companies to own directly somewhat less than half of their tonnage requirements, to time charter a quarter, often on long terms of five to ten years, and to voyage charter the balance. This arrangement allows companies flexibility in the choice of ships and manipulation of size of fleets as trading conditions fluctuate seasonally and alter from year to year, without burdening them with surplus shipping during periods of reduced demand for oil.

Operating very large crude carriers
The adoption of bigger vessels by tanker owners has enormously improved the productivity of oil-carrying ships. In 1963 every deadweight ton of active tanker shipping produced 28,000 ton-miles of oil haulage; by 1970 over 40,000 ton-miles was produced per deadweight ton.[6] Other improvements in tanker efficiency have been made, among the most important being the cathodic protection of tanks and hulls and new anti-corrosive coating materials, which allow

reductions in the thickness of plates and girders, and hence in the weight of the ship, and reduce docking and repair times for giant tankers to about twenty days per year. This off-hire time may be further reduced by the development of new paints, and underwater methods for cleaning the hull while in port, which will enable time between dry-docking to be extended and speed on passage maintained.

The improvements in efficiency, and the spate of shipbuilding which followed on the first Suez crisis, also produced adverse effects, for they contributed to an overtonnaging of the tanker trades between 1959 and 1966, and to a fall in freight rates. Many older vessels were scrapped, about one million tons was laid up, and another million tons transferred to the grain trade. The situation was saved, for shipowners, by the second Suez Canal crisis in the spring of 1967. Freight rates rose again in a few days, from £1·15 per ton for voyages previously made from the Persian Gulf to Europe via the canal, to a temporary height of £10 per ton for voyages around the Cape.

The high rates were short-lived, but world shipyards were once more filled with orders for even larger classes of ships with draughts well beyond the limits of the Suez Canal. The rate of adoption of the very large crude carrier (VLCC) may be appreciated from the fact that in the spring of 1967 there were 30 vessels of 200,000 dwt and over on order, four months later orders for this class of ship reached 90, and in the following year there were 125 ships of over 200,000 tons on order or being built. At the end of 1969 about 80% of the tanker deadweight tonnage on order comprised ships above 200,000 dwt. This latter size group will predominate in the carriage of crude oil in the future; it transported 115·3 million tons of oil in 1970,[7] against 46·4 million tons in 1969 and 10·4 million tons in 1968. Owners have clearly recognised that the economic gains to be made from the VLCC more than offset the longer voyages around the Cape, and that for the haulage of crude oil the Suez Canal can be dispensed with. Fig. 6 provides a clear indication of the relative merits of using the Suez Canal and the Cape route for various sizes of vessels.

Very large ships are, of course, extremely costly to build, and operating costs in absolute figures are high, consequently they must be kept working at maximum efficiency. Building and repair costs almost doubled in the period 1965–70, and several accidents on VLCCs in 1969 caused insurance rates to soar to three or four times the annual premiums anticipated in 1965. Taking ship and cargo together the annual insurance bill for a 200,000-ton tanker is, depending on the company's record, between £500,000 and £600,000. Total operating costs of such a vessel, including depreciation, amount to about £1·5 million per annum at 1969 prices. It will be appreciated from this that delays of VLCCs can be costly. The Shell Company estimated that a saving of £1m would have been achieved in 1968 had each of their ships saved one hour in port, and not all Shell tankers are in VLCC class.[8]

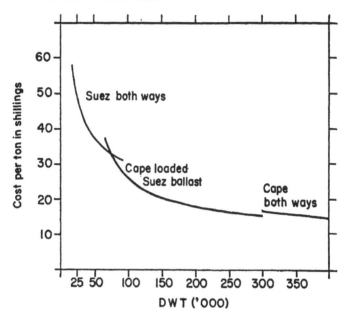

Fig. 6
Oil to Europe by size of ship and route. Source: based on McFadyan, F. S., *The Economics of Large Tankers*, Strathclyde lecture, 4 March 1968 (Shell Petroleum Co. Ltd, London)

The independent shipowner finds the adoption of the VLCC particularly risky. In the first place, he has to insure all his ships on the insurance market, unlike the major oil companies which are big enough to self-insure much of their fleet. If the independent owner accepts short-term charters then even minimum off-hire periods can prove extremely expensive, if he accepts very long-term charters he may be exposed to rising operating and insurance costs while earnings, as represented by charter rates, remain fixed. Some independent owners are, nevertheless, sufficiently risk-prone to work the spot charter market on the assumption that a crisis of one form or another will allow them to reap rich rewards, but the financial danger of having an unemployed VLCC on their hands is a deterrent to all but the boldest.

Both oil company and independent owners may in addition be faced with pollution liability. As we shall see below, this can involve immense sums of money in the event of spillage from a giant ship which has been in collision or has stranded; high indemnity or insurance rates must therefore be paid commensurate with size of ship. The increased tonnage of a ship means also loss of flexibility; not only does this result in the confining of ships to certain routes and ports but it restricts their choice of dry-docking and repair facilities in the world, and it could lead to the high costs of long tows to a suitable repair port should the vessel break down at sea. Experience of this with a new 208,000 dwt ship which had an explosion on board during 1969 showed that the cost of towage from South Africa to Japan, and then the repair costs, exceeded the original value of the ship. In such circumstances there may be the added costs arising from the disruption of refinery programmes when a giant ship drops out of schedule, and there will almost always be the costs of chartering alternative tonnage at short notice, which may mean very high charter rates for the oil company.

Another limiting factor to size of tanker is refinery throughput. It is necessary that refineries should be able to receive the cargoes of a VLCC without resorting to uneconomic storage. Very (Theoretically) tankers of 100,000 tons require a refinery capacity of about 3·5 million tons p.a., 200,000-ton vessels require a 10-million-ton refinery, and at 275,000 dwt the refinery throughput should be in the order of 15 million tons.[9]

The relationship between size of ship and refinery capacity actually depends on several factors. If a large ship is used to supply a refinery which serves a limited market it means a long time interval between shipments and hence more storage. The time between ships may be determined by the basic formula:

$$\frac{\text{capacity of ship}}{\text{tons used per day}}.$$

It is also necessary to allow for ship delays in the formula, and for the probability of the non-arrival of a ship, as well as for sudden increases in demand. The net result may be that substantial storage facilities are required.

Usually oil companies aim to minimise storage, and in an area such as north-west Europe, to which many ships operate, and where there are many refineries, there is normally little risk involved in holding minimum supplies. Here a refinery experiencing a shortage of crude oil through a ship dropping out of schedule can obtain assistance from other refineries. There are national differences, however. France in particular appears to hold greater strategic reserves of oil than does Britain.

In an area such as New Zealand it would be more of a risk, and expensive in storage costs, to use a 200,000-ton ship with a very long time interval between voyages. There are no alternative nearby suppliers to the New Zealand refinery at Whangarei and the country is distant from sources of crude. Consequently crude-carrying ships of between 50–60,000 dwt are favoured despite the long sea passage from the Persian Gulf. Another factor which may require the use of a smaller than optimum size of tanker is the need to blend crudes with different sulphur content from different regions, so that an occasional relatively small crude carrier may be introduced in supplying a high-capacity refinery at a deep-water site. Finally, an important prerequisite for the economic operation of VLCCs are berths which can be reached without undue tide or weather delays. So important is this that many oil companies are prepared to tolerate empty berths for long periods between ship arrivals if this ensures that the big vessels can get alongside without delays.

In order to attain the required throughput, speed of turnround, and maximum utilisation of the VLCC, oil companies have been altering again some of their refinery location criteria. A market-based refinery which also gives unimpeded access to a VLCC is not always possible. New refineries have therefore been developed at remote deep-water locations, and the installations of various oil companies have tended to group together in order to share deep water, or the capital costs of achieving this. Other methods of using the VLCC include multi-port discharge, whereby the vessel unloads at a sequence of refineries, moving from deep to less deep ports; or the ship may discharge at a deep-water terminal, possibly off-shore, and the oil is transported by pipeline to distant refinery locations. Yet another method is to lighten the ship at sea. By transferring about 25,000 tons of crude oil from a 200,000-ton tanker its draught can be reduced from 18·29 m (61 ft) to 16·15 m (53 ft), thus extending its choice of ports. An entrepôt may be used for a similar purpose, as at Bantry Bay in south-west Ireland. Here very deep-draughted ships of over 300,000 dwt are received and the crude oil is thereafter distributed by vessels with draughts suited to refinery ports in Europe.

The VLCC will undoubtedly go beyond the 500,000-ton class. How far beyond depends on the assessment by shipowners and oil companies of the economic and legal factors governing the operation of giant ships in the future. The high concentration of capital invested in a giant vessel, the refinery dependence on ship schedules which may be disrupted by the loss of a vessel, and the other risks inherent in say a million-ton tanker are very great; insurance companies as well as owners and governments are now increasingly sensitive to these. Other considerations in building to above about 320,000 tons include the need to fit two propellers which raises costs appreciably (Fig. 6). Then there is the factor of the Suez Canal.

A 250,000-ton ship, for example, could make a return voyage in ballast through the canal to the Persian Gulf. Many owners value this potential and the relative flexibility which the 250,000-ton class of ship gives in the choice of routes and ports, particularly as more of these are being made suitable to vessels of this size. Furthermore, one of the economic advantages of operating a ship of say 700,000 tons lies in the capital saving obtained from vast cargo tanks with few subdivisions; if, as appears likely, international legislation prescribes many more tank divisions as a safeguard against pollution arising from collision or stranding, then, when allied to increased insurance costs, the advantages of ships of enormous size are reduced.

Oil pollution and tankers

The question of pollution is a major one in connection with oil tankers. It should be remembered, however, that tankers are by no means the only ships contributing to the pollution of the seas and coastlines. The fact is, any oil-powered vessel may intentionally or unintentionally discharge oily waste into the sea, although the oil tanker is the ship which appears most likely to cause major pollution problems.

The main dangers of pollution from tankers arose with the shift of oil refining from the crude-oil-producing regions of the world to locations near the market areas. While refineries were located near the sources of crude it was largely refined non-persistent oils, such as gasoline, which moved on the high seas. If a spillage of this type of oil occurred it soon evaporated. When appreciable quantities of crude oil began to be carried, and when these built up to the huge amounts required by modern market-located refineries, the likelihood of the serious pollution of beaches, fisheries, and the ocean environment generally, increased.

There are three basic causes of pollution by oil tankers. First, the accidental spillage during cargo working, which is an infrequent occurrence and seldom leads to extensive pollution. Second, a ship may be involved in collision or become stranded, leading to catastrophic and widespread pollution. With the high concentration of shipping in certain areas these accidents are probable, but the risks can be enormously reduced, if not eliminated, by stringent routeing, efficient personnel on every vessel, and the shore control of dense traffic. The third cause stems from the discharge of oily water from the tanker during tank cleaning operations which take place *en route* between the discharging and loading ports. This source of pollution can be completely eliminated. Since 1955 there have been successive amendments to the Oil in Navigable Waters Act which prohibits the discharge of oil into certain areas of the sea. The restricted areas have grown progressively and the trend will undoubtedly culminate in the prohibition of the discharge of persistent oils anywhere in the ocean. The problems which still remain will be those of gaining adherence of all nations to the comprehensive restrictions, and the compliance of all ships on the high seas with these.

Most tankers already, in fact, observe strict measures in all waters. They allow their oily tank washings to settle in a separate compartment. The oil floats to the surface and the clean water is eventually dropped out of the tank leaving the oil and some water behind; the next cargo of crude is then simply loaded on top of this residue. However, in 1969 it was estimated that 20% of world tankers were still discharging about 600,000 tons of oily tank washings into the sea, but many of these ships are expected to adopt the 'load-on-top' system in the future.[10]

Tanker owners as well as taking safeguards against pollution have made provisions for the payment of compensation to national governments in the event of spillage, and especially in the case of catastrophic occurrences such as those which may result from collision or *Torrey Canyon*-type incidents. The companies involved in what is termed the Tanker Owners' Voluntary Agreement concerning Liability for Oil Pollution (TOVALOP) scheme controlled, in early 1971, over 70% of the persistent oils carried by sea transport. The fund, and a further supplement to it, then enabled payments of up to $30 million to be paid per pollution incident.[11]

There is no doubt that responsible tanker owners have stringent measures to avoid pollution, and they are, rightly, prepared to pay highly for any that occurs. Ships' officers are also very conscious of their responsibilities in this respect. But a number of owners, especially those obscured behind flags of convenience, will undoubtedly ignore the enforcement of many of these measures on their ships until detection methods improve. Such methods may extend to checks by infra red aerial photography which, as this becomes more sophisticated, will identify ships and oil slicks over a wide area.

Pipelines and tankers

The pipeline is another factor which could render difficult the decision on the size of ships to adopt for the future. In general, because of their high fixed costs and inflexible nature pipelines are seldom considered economic substitutes for the VLCC over long distances. Fig. 7 makes this apparent by showing that somewhere around the 500-mile distance ocean tankers reduce the costs of oil transport more rapidly than do pipelines. The almost horizontal

Oil tankers

trend of the curve representing pipeline costs is very significant in this respect. It indicates that operating costs per ton-mile do not diminish with distance, for the capital costs incurred in building and laying the pipeline predominate in the system of charging.[12]

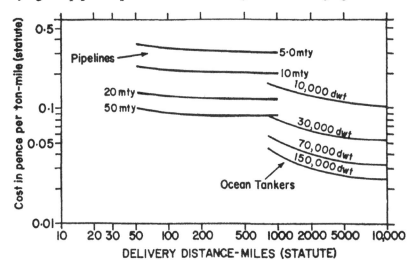

Fig. 7
Comparative costs of tankers and pipelines. Source: Hubbard, M., *The Economics of Transporting Oil to and within Europe* (London 1967)

The main problems in the pipeline/tanker controversy are strategic and political. There is no doubt that pipelines of relatively short length can reduce the demand for tankers by bridging interoceanic areas. Those crossing, and those planned to cross, the areas between the Persian Gulf/Red Sea region and the Mediterranean are of this category. Table 13 indicates the potential capacity of these pipelines, assuming they are completed and fully operative.

Table 13
CAPACITY OF MIDDLE EAST PIPELINES

Pipeline	Area	Capacity p.a. (mill. tons)
IPC	Kirkuk (Iraq)–Tripoli/Banias	65
Tapline	Saudi Arabia–Sidon	25
Eilash	Eilat–Ashkelon (Israel)	60
Sumed	Suez–Alexandria	60
Irtup	Iran–Turkey	60
		270

Source: Lambert Bros. (1971).

Of the pipelines shown in Table 13, the IPC line, Tapline and the Israeli Eilash line were operating in 1970, although not at full capacity. Out of the 335 million tons of Persian Gulf crude oil delivered to Europe in that year about 75 million tons was loaded at Mediterranean pipeline terminals and 260 million came via the Cape. If both the Sumed and Irtup lines are completed by 1975, and the crude-oil deliveries to Europe then amount to about 550 million tons, it would still leave the equivalent of the 1970 oil tonnage to be carried via the Cape by tanker.[13]

Much will depend on the level of royalty and transit charges for pipelines crossing national territories compared with tanker freight rates as to whether the pipeline systems can effectively compete with tankers and capture the additional build-up of oil. Some of the defects of pipelines may be illustrated by Table 14 in relation to tapline. When tanker freight rates were very high, at the closure of the Suez Canal in 1957, the pipeline was operating at maximum capacity and high transit charges were then imposed. When tanker rates dropped but the negotiated transit payments remained high the oil was diverted through the Persian Gulf port of Ras Tanura and the trans-Arabian line operated at reduced capacity. In 1967 tanker rates again soared but the pipeline flow was disrupted by political events. Unless, therefore, royalty and wayleave payments are low it is probable that the projected pipelines will have difficulty in competing with the VLCC. In any event it is most unlikely that oil companies will reduce their numbers of ships over 200,000 dwt since this would leave them vulnerable to pipeline closures as a result of political disruption, and as these ships are designed for the Cape route this will continue to be used. The same strategic thinking will, it might be noted, apply to the Suez Canal. Even if a combined pipeline and canal could supply all crude oil to Europe the oil companies would

Table 14

OIL FROM SAUDI ARABIA BY TAPLINE AND BY TANKER 1957–67
(mill. tons)

	Freight rate index	Mediterranean Tapline loading	Persian Gulf loading
1957	150	24	15
1961	48	15	32
1963	73	19	39
1965	66	22	56
1967	114 (political crisis)	16	86

Source: *ARAMKO Review* (1968).

nevertheless be unlikely to run-down their fleets of VLCCs suited to the Cape route. It is more likely that the canal will be enlarged to accommodate some of the VLCCs.

Where pipelines have been most successfully used over very great distances is in the United States. The products line from Louisiana to New Jersey covers 3,110 miles and eliminates much coastal shipping. A crude-oil pipeline extending from Alaska to the industrial regions of the eastern USA is possible, thus removing the necessity of tanker voyages through the north-west passage, but the costs would be extremely high. Another project for Alaskan oil involves a pipeline from Prudhoe Bay on the north coast 800 miles to Valdez on the south coast. Crude will be shipped from the latter terminal to US west coast refineries, and possibly to Japan. A short pipeline across the isthmus of central America with tanker connections at both ends would be suitable for delivery of Alaskan crude oil to the US east coast.

Pipelines are thus generally complementary to the activities of ships and allow the shortening of some tanker voyages. Tankers unloading at Trieste, Genoa and Lavera, for example, after loading in the Levant or arriving via the Suez Canal, can have their crude oil pumped overland to market-based refineries in the industrial areas of the upper Rhine. This amounts to a saving of about 3,500 miles in tanker transport to and from North Sea ports, and a further 300–400 miles of inland waterway carriage. But Hubbard suggests that by using very large tankers it would still be cheaper to deliver crude oil for southern Germany to North Sea ports via the Cape rather than through Mediterranean terminals and pipelines, and at times of low tanker freight rates this would most certainly apply.[14]

Patterns of oil trade
The main patterns of trade which result from developments in modern oil tankers are generalised by Figs 5 (p. 111) and 8. It is quite evident that with the enormous size of tankers the haulage of quantities of crude oil around the Cape will persist, whatever the developments in the Suez Canal route and trans-isthmus pipelines. Also, it is very likely that the activities of ships delivering refined products around the coasts of north-west Europe (Fig. 8) will continue to intensify as new refinery capacity tends to be grouped in more limited areas, and the organic chemical industries are also attracted to these locations by the availability of fuel oil, naphtha, liquid hydrocarbons and other products.

In many of the patterns of oil carriage it might also be noted that there is still some seasonality. In early winter there tends to be increased product-carrying activities in northern Europe especially

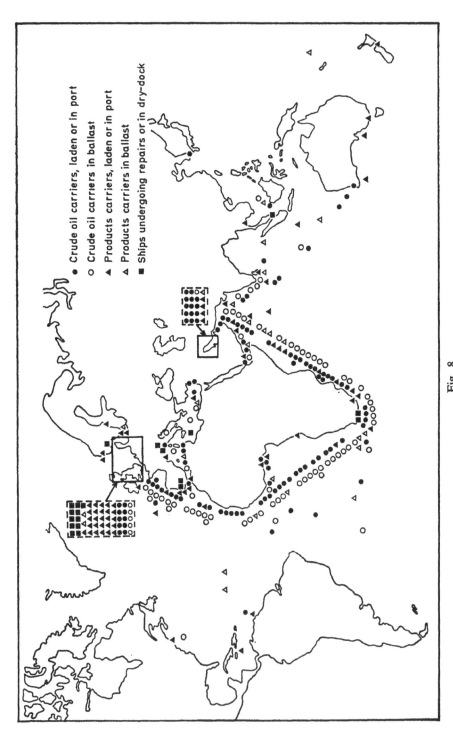

Fig. 8
Disposition of B.P. Group tankers (1 May 1969). Source: B.P. Tanker Co. Ltd.

in the delivery of heating oils and in the build-up of oil stocks at Baltic ports. In the northern summer petroleum companies take the opportunity of a lower demand for oil to dry-dock several of their ships, and many independent owners with clean-oil ships off-hire seek grain cargoes. These activities also mean variations in demand for crude-oil carriers. But it is difficult to show seasonal variations in the tanker trade over a long period in recent times due to the more pervasive effects of international crises. Fig. 9, however, contrasts

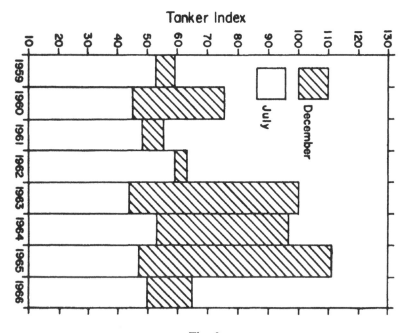

Fig. 9
Seasonal variations in tanker freight rates. Source: Petroleum Information Bureau.

the July and December tanker freight indices for the period 1959 to 1966, which was relatively crisis-free, and here a marked seasonality can be observed.

Finally, in considering future patterns of oil trade it might be noted that long-term shifts in sources of crude oil may come about as a result of pressures to reduce air pollution emanating from the products of high-sulphur crude. The Japanese oil market, for example, has shown the fastest rate of growth and a high demand for fuel oil in industry. Almost all the imported oil is Middle East crude which has a 40 to 50% fuel oil component but a high sulphur content. African crude by contrast has a low sulphur content but provides more gasoline than fuel oil. A reduction in pollution by a switch to more African crude would mean for Japan excess gasoline to dispose of in world trade. The USA is also faced with air pollution problems. Since 50% of the demand here is for gasoline and only a low requirement exists for fuel oil, there is thus considerable incentive to switch to African crude imports. Fig. 5 could therefore be modified, in the long term, by more substantial flows from Africa and more exchange of products between countries, as well as more crude oil from continental shelves.

Special tankers

There are a number of ships capable of carrying oil as an alternative to dry bulk cargoes. These combined vessels are discussed under the heading of bulk carriers (p. 136) and here only very specialised tankers built for liquid cargoes other than oil are briefly considered. The most important of these are the liquid-gas tankers. By liquefying natural gas (LNG) and petroleum gas (LPG) it is possible to transport such products by sea in a highly concentrated form (LNG at 1/600th of the gas volume) and to return them to a gaseous state after discharging at special terminals.

The liquefied natural gases carried by tankers comprise methane plus ethane, propane and butane. Methane has a boiling point of $-162°C$ and is carried in special spherical or prismatic tanks. Several LNG vessels, of 5,800 dwt, trade regularly between the port of Arzeu in Algeria and Canvey Island on the Thames with methane from the Sahara. This particular linkage by small ships has been challenged by piped North Sea natural gas. On the other hand some large liquid-gas tankers of over 40,000 dwt are increasingly being employed for longer hauls from Venezuela to the USA, from the Persian Gulf to Japan and from Brunei to Japan. Other developments include even bigger liquid-gas carriers, costing between £25 and £30 million each.

Tanker transport of liquefied petroleum gas has also increased in recent years. This trade includes propane, propylene and butane which occupies 1/250th of the gaseous volume when in liquid form under pressure or refrigeration. As well as occurring in natural fields LPG is obtainable from the refining of crude oil so that the places of origin of these cargoes are quite widespread. The total fleet of all liquid-gas tankers has grown from 10 vessels in 1955 to 292 in 1969. This reflects the increased demand for the product as a source of hydrogen, acetylene and ammonia.[16] Costs of transport on a calorific basis may be slightly higher than for oil, but with the rapid technological advances in this field the haulage of liquid gas by very big ships will undoubtedly increase. The low pollutant nature of liquid gas as a fuel makes it particularly attractive for industries in Japan, USA and parts of Europe. The principal fleets of liquid-gas carriers are shown in Table 15 for 1968.

Table 15
PRINCIPAL FLEETS OF LIQUID GAS AND CHEMICAL CARRIERS 1968

	Liquid gas		Chemicals	
	No. ships	m^3	No. ships	dwt
Japan	62	418,086	—	—
Norway	33	189,766	43	596,138
France	16	136,386	—	—
UK	15	102,509	29	201,639
Sweden	9	66,300	—	—
Liberia	10	—	10	151,295
USA	—	—	15	266,312

Source: *Liquid Gas Carrier Register* (H. Clarkson and Co. 1969) and UK Chamber of Shipping.

The chemical industry is likewise employing specially constructed tankers, the fleets of which are also shown in Table 15. These carry cargoes such as molten phosphorus, sulphuric acid, anhydrous ammonia and ethylene. There are economies of specialisation and scale involved in the adoption of tankers by the chemical industry, for it allows industrial plants to concentrate on one or two products and have these distributed widely. There are also economies in shipping, for a cargo of anhydrous ammonia contains the nitrogen fertiliser equivalent of four dry cargo ships of similar deadweight. Sulphur can be carried in liquid form and, as with other solids, when molten this means more concentration and space saving. In addition, liquefied sulphur eliminates losses of cargo through wind, rain and contamination, which inevitably occur when chemicals are transported in lump or powdered form. In almost all cases liquid cargoes can be loaded and discharged at over 500 tons per hour compared with 1,000 tons per day on tramp ships.

Some tankers are built for the trade in molasses, and a more limited application of tankers may be found in the carriage of wine from Cyprus and Portugal to Britain, and orange juice in the trade from Florida to New York. All of these cargoes require special ships and, to various degrees, close integration with shore installations at their places of origin and destination.

1. *Fearnley and Egers Review 1971* (Oslo 1972), p. 8.
2. King, G. A. B., *Tanker Practice* (Maritime Press Ltd 1968) provides detailed descriptions of tankers.

3. Davis and Newman Ltd, *Analysis of World Tanker Tonnage* (London July 1971).

4. Sturmey, S. G., *British Shipping and World Competition* (London 1962).

5. Odell, Peter R., *Oil and World Power* (Harmondsworth 1970), p. 159.

6. *B.P. Statistical Review* (1969) and data (Petroleum Information Bureau).

7. Fearnley and Egers, *Large Tankers* (Oslo January 1970-1).

8. Kirby, J. H., 'Large Tankers and their Problems', *Shipping World and Shipbuilder* (July 1969), pp. 940-1.

9. Westinform Report. *Demand for Giant Tankers 1970-1975* (September 1968).

10. B.P. *Fleet News 15* (January 1971), p. 8.

11. TOVALOP. *Tanker Owners Voluntary Agreement Covering Liability for Oil Pollution* (London 1969).

12. 'Pipelines versus Tankers', *Norwegian Shipping News 4* (1968), pp. 130-1.

13. Lambert Bros. (Shipping) Ltd, *Middle East Pipelines: The Impact of Crude Oil Pipelines in the Middle East on Demand for Shipping in 1975* (London 1971).

14. Hubbard, M., *The Economics of Transporting Oil to and within Europe* (London 1967).

15. Wheeler, R. P., 'Ocean Transportation for Liquid Gas' (Conch Methane Services) *Financial Times* (29 July 1968), p. 28, and Gray, R. C., and Johnson, C., 'The Design and Construction of Liquefied Gas Carriers', *Transactions of the North East Coast Institution of Engineers and Shipbuilders* (February 1971), pp. 69-79.

8

BULK CARRIER AND UNITISED SHIPPING

As in the case of the oil tanker, the giant bulk ore carrier has resulted in very substantial reductions in the costs of transport; and the lower average delivered price of ore moved by bulk carrier has, over a period of years, exerted a force sufficient to influence locational decisions in the iron and steel industry in many parts of the world.

In the case of unitised shipping the concept, and some of the methods, of handling cargoes in bulk have been applied to shipments of heterogeneous commodities. This has involved not only advances in maritime technology but new approaches to transportation. Shipping has become recognised by shipowners and shippers alike as one link in a through-transport system extending from producers to consumers. This total-cost concept allows a high degree of investment to be made at certain sectors of the transport system. and higher charges to be levied there if necessary, if by so doing total costs are reduced.

The maritime industry in advancing to giant bulk carriers and unit loads has thus crossed the threshold of simply improving on conventional vessels to adopting ships quantitatively and qualitatively different from those in the past. With the increased scale of activities, and the standardised units involved, the industry has thereby entered an era of mass production in transport services.

Dry cargo bulk carriers

For statistical purposes the bulk carrier may be taken as any ocean-going, single-decked, dry cargo vessel of over 18,000 dwt. This obviously includes some ships discussed previously under the heading of tramps. Only in the case of specialised bulk carriers, giant vessels, and combined carriers are there any distinctive differences from the tramp; the figures given in this section under the heading of 'other bulk carriers' cannot therefore be directly compared with those in Chapter 6.

The carriage of commodities by large bulk vessels has been on the increase in quantity and range of goods throughout the post Second World War period. The practice has been encouraged by the obvious economies of bulking; it is, for example, clearly cheaper in transport terms to carry 50,000 tons to a destination in one large vessel rather than on ten smaller ships. The shipping of such huge quantities at the one time has been made possible by the capacity of many industries to cope with higher throughputs of materials, by the activities of multi-national companies with their raw material bases and processing plants at widely separate places, and by the formation of producers' and shippers' organisations which accumulate commodities for shipment. The trend to bulk carriage has also been stimulated by the introduction of more mechanical handling equipment at the ports to reduce manual labour, and by the technological advances in shipping.

Many commodities are now moving in bulk form, but a few predominate, namely iron ore, grain, coal, bauxite and alumina, and phosphate. If we assume that conventional tramp ships are below 18,000 dwt, then the increasing role of bulk carriers may be appreciated from Table 16.

Table 16

WORLD SEABORNE TRADE OF MAIN BULK COMMODITIES 1960 and 1969
(mill. tons)

	1960		1969	
	Total (all ships)	By bulk carrier (above 18,000 dwt)	Total (all ships)	By bulk carrier (above 18,000 dwt)
Iron ore	101	31	214	181
Grain	46	1	60	36
Coal	46	3	83	60
Bauxite and alumina	17	3	30	19
Phosphate	18	—	32	12
Other bulk commodities	not available	not available	not available	61
	228	38	419	369

Source: Fearnley and Egers Chartering Co. Ltd (1970).

The other commodities commonly carried by bulk vessels include manganese ore, ilmenite, chrome ores, coke, scrap iron, pig iron, steel products, timber, wood products, sugar, soya beans, salt, fertilisers, cement, sulphur, gypsum and pyrites. The increased productivity in bulk shipping is indicated by a rise in transport performance per deadweight ton by over 30% between 1960 and 1969.[1]

A bulk carrier need not, of course, be of giant size. It may be built for the carriage of cargoes the economic loads of which when related to industrial requirements are relatively small. It may also be required to load at ports with restricted access and where improvements to allow bigger ships might prove uneconomic in relation to the likely duration of mineral production in the hinterland. It should be recalled in this respect that, unlike oil, iron ore and other heavy minerals cannot as yet be loaded far off-shore in deep water without costly equipment, although progress has been made in the loading of crushed iron ore in wet slurry form which can be pumped from and to inland locations through pipelines. This has led to the use of off-shore terminals similar to tankers' moorings.

The main developments of bulk carriers have in fact taken place in connection with shipments of iron ore for the world steel industry. We can trace the evolution of the huge bulk carrier through the growth in the postwar iron ore trades, and in the 1970s can readily observe the impact of these ships on the development and location of the iron and steel industry.

It is quite remarkable to recall that in 1937 the total seaborne trade of all mineral ores amounted to only 25 million tons. This was hauled by tramp ships of around 5,000 dwt. The main routes were from Scandinavia, north Africa, Spain, Newfoundland and west Africa to Europe and Britain, from Chile to east-coast USA, and from Indo-China and Australia to Japan.[2] Most of the iron ore then used in the USA, Britain and Europe was obtained from home deposits, although by the outbreak of the Second World War Britain was importing almost one-third the blast-furnace charge of iron ores.

At the end of the Second World War suitable ferric and non-ferric industrial minerals appeared seriously depleted in many of the manufacturing countries of Europe. Despite the technical advances which soon made possible the use of lower-grade home supplies, there was a growing dependence on foreign ores. In 1946, of the 19 million tons of ore required by the British steel industry about 6 million was imported. By 1963 home ores and foreign ores were used in equal amounts in Britain, but from then on imports predominated. Many fresh deposits were discovered in Africa, Canada, South America and Australia, which were often of high grade and could be delivered at European steel works at a lower price per ton of iron content than could some home ores. The Benson Report (1966) showed that, in Britain in 1957, home ores and imported ores had parity of price, but by 1966 home ores had increased in price by 28% compared with a reduction of 30% in the price of imported ores.[3] As a result of these price trends favouring imports, and the expanded market for steel, the ton-miles in ore transportation by sea, have, according to Fearnley and Egers statistics, increased from 264 thousand million in 1960 to 919 thousand million in 1969 and 1,100 thousand million in 1971, representing an enormous growth in the dependence of the world steel industries on ocean transport (Fig. 10).

The increased use of iron ore came as the pace of the industrial economies accelerated in the years from the 1950s onwards, first in response to postwar reconstruction, then for rearmament, especially during the Korean war, and finally with increased consumer demand and major constructional projects in relation to urbanisation and transportation. The iron and steel industry by the mid-1950s included some massive strip-mills and multi-product integrated plants, especially in Europe and Japan, where reconstruction and rationalisation had proceeded rapidly. With the greater throughput of these plants (several of over 4 million tons per annum) steel makers could no longer rely on deliveries of ore by tramp ships chartered on the open market. They required regularity of supply and stability of freight rates; to obtain this some steel companies built their own ships, many others offered shipowners long-term contracts for suitable iron-ore carriers.

The new high-capacity steel plants could, if they were sited with access to deep water, accept greater and therefore cheaper ship-loads of ore as reductions in the landed price of distant ores was largely a function of ship size. It is significant for the success of Japanese industry that most of the larger plants in Japan were from their inception coastally sited. Here there were no inherited ore or coalfield-located steel industries. What therefore had been a disadvantage was now a considerable advantage, for Japan benefited almost immediately from cheaper overseas ores. It was difficult for the British steel industry to take advantage of the economies of the bulk carrier, since much of their steel-making capacity was located in older inland industrial areas; or, as in south Wales, at ports with access restricted to 22,000-ton ships. Many of the traditional inland areas no longer held all the advantages they once possessed. Greater efficiency in the use of coke in steel production had diminished the pull of the coalfields, some cheaper coke was being imported and oil fuel was increasingly substituted for fuel and power. These factors gave higher locational values to deep-water coastal sites

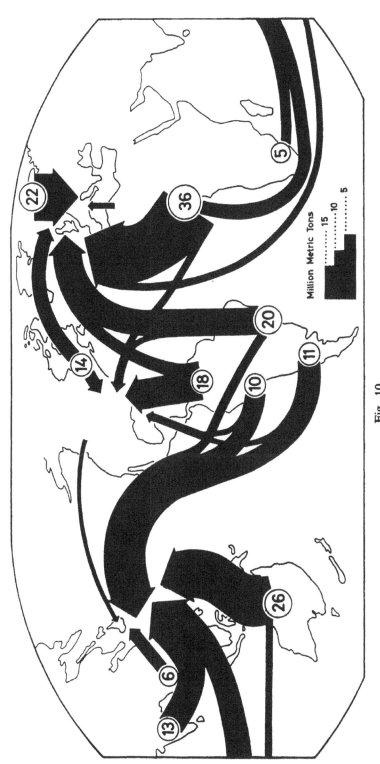

Fig. 10
Iron-ore trades (1969). Source: Fearnley and Egers Chartering Co. Ltd.

which could be used by bulk carriers delivering overseas ores. There was, however, so much industrial and social capital built around steelworks in traditional locations that any spatial change could proceed only slowly. As will be noted again in this chapter, the delay in the ability to use the giant ship had serious consequences for British steel throughout the 1960s.

The 1960s saw the expansion of the world bulk fleet from less than 7 million to almost 70 million dwt. The types of bulk carriers also changed in the latter part of this decade from a predominance of specialised iron-ore carriers to more flexible vessels. Many of these new ships are capable of carrying ore, oil, or other commodities on alternative legs of voyages. The various types of bulk vessels are discussed below.

Types of bulk carriers

Many of the ships built from the late 1950s onwards were single-purpose ore carriers. That is they were large heavily constructed vessels built solely for the carriage of iron ore. This cargo has a low stowage factor of about 0·3679 to 0·5660 m^3 (13 to 20 cubic ft) to the ton; consequently when conventional tramps were employed there was a high proportion of empty space in the holds, and often a low centre of gravity, causing the ships to roll violently. On bulk ore carriers this empty space is reduced by fitting deep double-bottom and side tanks which can be utilised for water ballast on voyages to ore-loading terminals, and the height of tank above the bottom allows ore to be stowed higher in the ships, thus raising the centre of gravity and reducing racking stresses due to rolling.

Bulk ore carriers of the single-purpose type spend about half their time in ballast, as they generally operate shuttle services between ore terminals and deep-water industrial ports. Whether owned by steel companies or operated under long-term charter, such vessels are frequently tailor-made for particular routes and ports. Consequently, they are somewhat inflexible and can seldom obtain return cargoes in the direction of ore-loading terminals. The main attributes of the bulk ore carriers are their relative simplicity and cheapness of construction and their speed of turnround in the specialised ports for which they are designed. Loading rates of at least 40,000 tons per day are expected in the ports handling very large vessels of this type.

From the single-purpose ore carrier there has developed a dual-purpose combined carrier aimed at minimising ballast voyages. On this vessel the double-bottom and wing tanks are adapted to the carriage of oil. This involves higher initial expenditure on pipelines and pumps as well as in design, but it means the possibility of complementary trading relationships over certain routes by the haulage of ore in one direction and oil in the other, or the use of the ships in triangular patterns of trade of which one leg is in ballast. It can also mean more flexibility in the employment of the ships by their ability to move from ore to oil haulage on a seasonal basis, or according to unexpected demands in one or other of these trades.

An even more flexible type of bulk carrier has evolved since 1964. This is the triple-purpose ore/bulk/oil ship. The OBO vessel can carry ore or oil, or other bulk cargoes, either in full loads throughout the ship or as parcels of commodities in separate compartments. In addition there are several other types of bulk carriers, including the universal bulk ships (UBS) which are simply large single-decked vessels of 18,000 to 30,000 dwt used in a wide variety of trades and often fitted with cranes. Bulk carriers with self-unloading equipment are in use on the Great Lakes and have been employed on the transatlantic iron-ore trades. They are fitted with hold conveyor belts, or bucket systems, and can discharge at rates of 20,000 tons per hour. The high capital cost of shipboard equipment will restrict this type of vessel to short hauls. Bulk coal carriers of 90,000 dwt have been employed on the very long hauls between Hampton Roads (US) and Japan via the Cape of Good Hope and for the Canadian west coast to Japan route, and bulk timber carriers of around 30,000 dwt built with long hatches and special cargo equipment to handle unit loads of timber now predominate in the trade from British Columbia to Europe, via Panama. The latter vessels enable timber unloading rates to reach over 20 standards (about 60 tons) per gang-hour compared with four or five standards using conventional ships. Other specialised ships include cement carriers, bulk sugar carriers, and smaller bulk vessels for the transport of minerals such as zircon and lead. There are also developments in dual-purpose bulk/car carriers fitted with folding hold decks employed in the carriage of motor vehicles outwards, especially from Germany and Japan, and grain, ore, or coal on the return voyages. Yet another advance in bulk handling is the slurry carrier. This is akin to an oil tanker in that pulverised bulk materials are mixed with water and loaded and discharged through pipelines. The water is decanted after loading and the solids compact during the voyage, the decompacting is achieved by water jets during discharging. These ships were at an early stage of use in 1971.

Of all the above ships the combined carriers of from 40,000 to 120,000 dwt have proved particularly attractive to shipowners who are seeking charters, for they have enabled them to move between the ore, oil and other bulk markets as trading conditions altered. In 1966, for example, when the tanker freight rates were low, 64% of the total transport capacity of combined carriers was devoted to ore, but in 1967 when there were record highs in tanker freight rates combined carriers devoted only 13% of their transport capacity to iron ore. These vessels also carried coal and grain during this period but in relatively small quantities.

In order to work the market with combined carriers and take full advantage of freight-rate fluctuations the shipowner must avoid having his vessel tied to long-term charters. It will be seen from Table 17 that the Norwegians, who depend almost exclusively on charters, have a high preponderance of these flexible vessels. As well as enabling shipowners to switch from oil to dry bulk cargoes, and vice versa, with changes in freight rates the combined carrier also allows the minimising of ballast voyages. In a sense the operators of combined vessels are attempting to recapture some of the advantages of conventional tramp ships, with the added attraction of higher capacity and the facility for oil haulage. This is reflected in their patterns of trade, an example of which is shown in Table 17.

Table 17
PRINCIPAL BULK CARRIERS BY FLAG 1970
(000 dwt)

	Bulk/ore	*Ore/oil*	*OBO*	*Other bulk*
Liberia	1,620	2,726	1,853	11,139
Japan	3,866	2,461	465	5,956
Norway	403	607	2,377	8,160
UK	858	152	479	505
Italy	158	392	338	2,366
World	8,389	8,035	6,355	48,387

Source: Fearnley and Egers Chartering Co. Ltd (1970).

Patterns of trade and freight rates

The various types of bulk carriers have different patterns of trade. The combined ore/oil ships, for example, were developed to a limited extent as early as the 1920s for the carriage of oil to Scandinavian ore terminals and iron ore outwards. They later became the most obvious type of ship for companies engaged in the Labrador iron-ore trade, as the closed season for ore exports from this area coincides with the high winter demand for oil transport to which they could transfer. In more recent years complementary patterns of trade using both combined ore/oil and OBO ships have developed on a world-wide basis. This may be appreciated from the voyage routeings listed in Table 18.

Table 18

VOYAGES OF ORE/OIL AND OBO BULK CARRIERS

Origin	Cargo	Destn.	Cargo	Destn.	Cargo	Destn.	Cargo	Destn.	Cargo	Destn.
USA	Grain	India	Ballast	P. Gulf	Oil	USA				
USA	Grain	India	Ballast	P. Gulf	Oil	Far East	Ballast	British Columbia	Grain	Europe
USA	Coal	Europe	Ballast	Liberia	Ore	USA				
USA	Coal	Japan	Ballast	Chile	Ore	USA				
USA	Ore	Japan	Ballast	S.-E. Asia	Oil	USA				
P. Gulf	Oil	Medi.	Ballast	West Africa	Ore	Europe				
P. Gulf	Oil	W. USA	Ore	Japan	Ballast	P. Gulf	Oil	Europe		

Source: Westinform (1969).

Such patterns of trade have proved highly economic, for example the last voyage listed in Table 18 occupies 30% less time than if the tonnage were used on shuttle services. In this case the saving in operating costs is 25% when employing a vessel between 70,000 and 80,000 dwt.[4] Another example of complementary trading which has reduced total costs may be cited from the St Lawrence. Here combined carriers deliver grain from the Great Lakes to ocean-loading terminals such as Baie Comeau, proceed in ballast to Sept Îles and there load iron ore for return to lakeside steel ports.

Giant ships have tended to dominate the ore trades. This is not the case with grain, the third most important dry bulk cargo by tonnage. Outside of the Great Lakes few vessels have been built solely for the carriage of grain. This is understandable for, as Fig. 11 (p. 142) indicates, the main route for this commodity is from the US east coast to Europe, a moderate distance compared with ore haulage, and a route over which there is almost always considerable tramp ship, cargo liner and tanker traffic capable of carrying grain at low rates when necessary. In addition the demand for grain fluctuates with harvest conditions in importing countries so that there is less security for specially constructed grain ships and few long-term charters. This makes the grain trade eminently suited to tramp ships. However, combined carriers and bulk/car vessels have, along with other bulk ships, tended to raise the average tonnage of vessels carrying grain in recent years.

Another factor which is now tending to raise the size of ships engaged in the grain trades is the attempt by North American exporters' organisations to reduce transport costs in the interests of producers. This they are doing by concentrating grain shipments through high-capacity loading terminals at the lower end of the St Lawrence. Ports such as Baie Comeau and Port Cartier are open for longer in the year and are closer to Europe than upstream loading places. With loading rates of 3,000 to 4,000 tons per hour it is possible to employ grain carriers of 100,000 dwt from North America and Australian ports. In order to do so changes must take place in the unloading areas and patterns of marketing. In Britain, for example, many ports have installations for the importing and storage of hard grain used in the production of bread flour. The flour mills in these port areas are, in turn, closely related to regional markets so that the whole system is geared to regular shipments by relatively small vessels of 10,000 to 25,000 dwt. In Rotterdam by contrast the port-located flour mills serve a populous hinterland of industrial cities, and the handling and storage facilities can accept cargoes of over 100,000 tons at the one time. With the economies obtained from using ships of this size on long voyages from Australia it has proved economic to tranship grain from Rotterdam to Britain. The trend in most countries is to concentrate grain shipments into fewer ports containing high-capacity stores, and for the milling industry to concentrate and expand capacity to a national rather than regional orientation. This is apparent at London, Liverpool and Hamburg. Bigger ships will also be adopted in the future for the haulage of maize used in the breakfast-food industry which will likewise concentrate in a few ports for distribution to national markets.

Another type of large bulk ship which has lowered the freight rates for grain is the tanker. Grain has some of the characteristics of a liquid in that it flows under gravity or suction; this has enabled oil tankers of 30,000 to 40,000 dwt to enter the grain trade during the summer lag in oil haulage, or whenever there has been an oversupply of tankers in relation to the demand for oil. When this latter situation has coincided with a fall-off in the demand for grain as a result of favourable harvests in importing areas, grain freight rates have been driven to low levels. In the autumn of 1966, for example, bulk carriers received high rates of $6·75 per ton for grain on the US Gulf to continent trade. In 1968 there were many oil tankers unemployed, good grain harvests in India, and surpluses of grain in the main producing areas. Rather than lay up their ships, tanker owners were accepting $2·70 f.i.o. on the US east coast to continent trade and they set a level for the earnings of other vessels. The main disadvantage of tanker tonnage in the grain trade is the protracted time of discharging due to restricted tank openings, and the regulations in some countries against the use of grain carried by tanker for human consumption. The long-haul grain trades are shown in Fig. 11 for 1969.

The bulk carriers have been important also in the revival of the world coal trade, for they have enabled the Japanese steel industry to draw widely on many sources. In the haulage of coal between Hampton Roads (US) and Japan 90,000 dwt ships have operated profitably at $5 to $6 per ton via the Cape of Good Hope compared with $9 to $10 necessary for 16,000 dwt vessels proceeding by the Panama Canal route. Bulk carriers have as a result increased their participation in the world coal trades from 15% in 1960 to 72% in 1969. These coal carriers are frequently delivering coking coal for the steel industry and therefore use the same deep-water discharging ports as ore carriers. To take advantage of this it has been necessary to construct new deep-water ports in the coal-loading areas. The coal trade between western Canada and Japan is for these reasons likely to see the operation of very large vessels, possibly acting to the detriment of the traditional American east coast loading areas from which ships are faced with a Panama Canal passage or

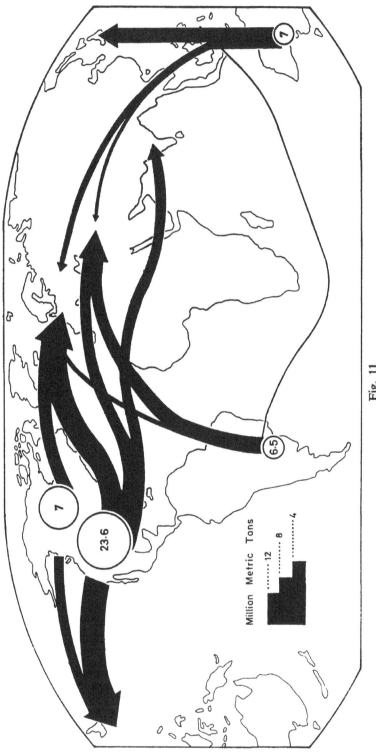

Fig. 11
Grain trades (1969). Source: Fearnley and Egers Chartering Co. Ltd.

an exceptionally long haul around the Cape of Good Hope to Japan.

Freight rates in the softwood trades have been lowered with the use of the 30,000 dwt timber carriers already referred to. It is sufficient to point out that at 1968 prices a bulk timber carrier with six hatches and travelling gantry cranes could deliver timber from British Columbia to Europe at half the cost of a 10,000-ton vessel.[5] In the tropical hardwood trades from west Africa liner vessels still predominate. Timber is loaded in relatively small quantities at many ports in west Africa for discharging in the ports of several countries of north-west Europe. In south-east Asia greater quantities of hardwoods are shipped from fewer points to concentrated markets in Japan and Korea. Purpose-built bulk timber carriers with special handling equipment have been introduced by Japanese companies into this trade.

Geographical impact of bulk carriers

Many of the geographical implications of the trends to the carriage of materials by giant ships have been touched on above. They affect numerous aspects of production and distribution and hence relationships between areas. For example, because of their low operating costs per deadweight ton bulk carriers have lowered the acceptable level of freight rates on many commodities. These ships now operate rates more than 20% below what was once the lay-up level for tramp vessels and this has proved of great benefit to the world economy, although for shipowners lower costs have not meant higher profits, merely the ability to compete at lower freight rates.[6]

As well as stimulating trade in low-value materials, the general lowering of transport costs has allowed some industries to adopt a more transport intensive policy in their industrial strategy; by, for example, moving semi-processed materials widely between plants in different countries. This has allowed geographical variations in the cost of labour, land, power and levels of taxation to be exploited. Linkages by sea transport between vertically integrated plants over several regions of the world were already noted in connection with the oil and chemical industries, they can be observed also in the production of steel, cars, and aluminium to which bulk carriers contribute the transport factor.

The aluminium industry, although still tending to change its spatial relationships quite rapidly, provides interesting examples of dispersed industrial locations based on variations in factor costs. In Australia, for example, bauxite from Wiepa on the Cape York peninsula is transported 1,300 miles by sea to the port of Gladstone in Queensland. There local coal is used for calcinating the ore to produce the alumina. This is then carried to coastal smelters at Bell Bay, Tasmania, and to New Zealand, where industries based on hydro-electricity carry out the electrolytic reduction of alumina to produce aluminium metal. Consignments of the latter are returned to foil-making plants in New South Wales and also shipped to manufacturing industries in Japan. In this example movement is from an area of tropical bauxite deposits to one in a temperate zone where hydro-electricity can be cheaply produced from snow melt. The intermediate stage, the production of alumina, is based on coal resources and caustic soda located between the bauxite and the source of hydro-power. Low freight rates by specialised transport allows the distance factor between these inputs to be radically reduced.

Under the EEC integrated industrial structure similar linked processes will expand. Increased tonnages of imported bauxite will be processed in Sardinia, for example, to be followed by wide distribution to other member countries as alumina. The most impressive example of such process relationships must be the transport of Australian bauxite to Iceland where hydro-power converts it to alumina for return to Europe. By contrast the Swiss aluminium industry, which at one time depended on the French deposits at La Baux, has ceased to figure as a major producer due partly to lack of direct access to ocean bulk carriers.

When major industrial linkages are made between countries which are not politically as well as economically integrated, bilateral trade agreements become necessary. As well as ensuring regularity of raw material flow and freight rate stability for the industry many agreements have been designed to protect the raw material producing areas. The Japanese steel industry draws half its coal requirements from Australia and this represents 90% of Australia's coal exports. These have increased from 3 million tons in 1961 to 14 million tons in 1969 in response to Japanese needs. Similarly, almost 90% of the iron-ore export of Australia in 1969 was destined for Japan. In both cases Australia's dependence on a single market has required the security of long-term agreements in the form of bilateral trade pacts. The effect of these arrangements is to regularise to an even greater extent world flow patterns.

The impact of giant ships on the geography of countries is nowhere more strikingly illustrated than in Japan. In order to overcome the double disadvantages of a paucity of industrial minerals and remoteness from sources of supply the Japanese have been foremost in the building and chartering of bulk ships. They thus found it necessary very early in bulk development to deepen and extend their ports to accommodate vessels of over 100,000 dwt. This allowed giant ore carriers to haul cargoes from distant sources at an economic rate. On a round voyage of 20,000 miles, with the outward leg in ballast, the transport costs per ton of ore on a 30,000 dwt vessel would be about £2·50; by using a 70,000 tonner this would reduce to £1·50, and probably to about £1·20 on a 100,000 dwt ship, provided always the rates of loading and unloading do not reduce with size of ship these economies continue at a diminishing rate. The new port installations built in Japan, Brazil, Australia and Africa in the 1960s have in fact kept cargo handling well above 40,000 tons per day; indeed the new port of Tubaras in Brazil was, in 1971, building loading facilities with a capacity of 30,000 tons per hour.[7] The resulting reductions in the economic distance between Japan and world ore deposits enabled Japanese steelmakers to import ore at lower costs per ton from west Africa and South America than British steelworks could obtain from Scandinavian sources. The consequent reduction in steel prices as a result of cheaper ores, and the economies of scale from greater throughputs at large port located plants, had a multiplier effect on the whole of the Japanese economy. It assisted, for example, the car and shipbuilding industries to produce their finished products at lower prices.

The critical factor in the use of giant ships is clearly the depths at, or approaching, the loading terminals or steelworks. Unlike oil tankers, iron-ore ships require discharging berths which are solid structures with heavy and expensive handling equipment. These should be near the blast furnaces as transhipment costs could reduce the economies of ocean transport. Attempts to overcome the access problems have included dredging programmes and the building of exceptionally broad-beamed vessels to minimise draught. But if the port remains restricted by depth, or by lock dimensions, to relatively small vessels, it means that port forelands for ore supplies are likewise limited in extent, or else the industry has to employ a less than economic size of vessel for long distance deliveries.

The relationships in this respect between size of ship and distance from Britain are, roughly: over a 4,000-mile round voyage a 40,000 to 60,000 dwt ship proves most economical, while a round voyage of 6,000 miles would require a 70,000 or 80,000 tonner for optimum working.[8] The actual optimum transport distances would, of course, also depend on the age and speed of the ship and the rates of cargo handling, but they would probably be of the above order. Clearly, if a relatively small ship is used over a long distance shipping costs per ton will be higher, and so usually will charter rates, compared with a big ship; in September 1968, for example, a 70,000-ton vessel was obtained at $4·85 per ton for ore from Port Etienne (west Africa) to Japan, but from nearby Port Noite ore on a 20,000-ton vessel cost $9·65 to ship to Japan.[9] Countries which cannot offer ore-discharging ports with depths to match loading ports, and which use vessels not in accord with size-distance relationships, find that not only are their costs of ore transport more expensive than those of their competitors, but their bargaining power in relation to nearby ore producers is also reduced.

The above factors have, in fact, been the basic differences between British steel production on the one hand and Japanese and some continental producers on the other. Restricted access to British steel ports has meant that the industry has, until recently, been unable to take full advantage of cheaper ores being exported from Africa and South America. In 1966, for example, only 10% of British ore imports were delivered by ships above 25,000 tons, whereas almost 70% of the ore delivered to ports on the continent of Europe came in vessels above this size.

The seriousness of the position for Britain may be better appreciated from the calculation that in the carriage of ore to the deepest of United Kingdom ore terminals in 1967 (the berth for 35,000 dwt on the Tyne) an extra 75p per ton was added to the landed price of ore brought from 5,000 miles compared with that landed at deeper continental ports by bigger ships. Furthermore another 35p to 40p per ton was added for the haulage of the British imported ore by rail from the Port of Tyne to blast furnaces at the inland steelworks of Consett in County Durham. Since 1·7 tons of ore is required to produce 1 ton of steel this meant that an additional £2 per ton was added to the cost of British steel compared with that of some continental rivals.[10] Due to the many alternative suppliers available such increases could not easily be passed on to overseas buyers of steel, a matter of some concern since about 20% of British steel is exported. The price differential also placed British shipbuilding at a competitive disadvantage, since the price of steel is a principal cost in the shipbuilding industry.

The economies which giant bulk carriers can offer are thus the fundamental reasons for the extension of the steel plant and deepening of the ore terminal at Port Talbot in south Wales to accommodate vessels of over 100,000 dwt. Precisely the same developments are taking place on the Humber, Clyde, Tees, and at locations on the seaboard of north-west Europe, the Mediterranean, the eastern seaboard of the USA, and in Japan. One major development is thus clearly towards a small number of highly integrated steelworks at coastal locations for the reception of bulk carriers.

The trend in world trade will be towards further movement of commodities in bulk. Allied to this, more elaborate types of carriers are being designed to reduce the costs of transport even further by the carriage of semi-processed materials instead of raw materials, thus eliminating the haulage of waste. The siting of semi-processing plants on long-life ore deposits has been made more possible by the joint financing of mining by multi-national companies in conjunction with governments. Bulk carriers hauling copper or iron concentrate, pelletised ores, or ore slurry which can be pumped to and from the ship, will undoubtedly increase in number. The slurrying process has made possible the exploitation of low-grade ores in difficult inland areas in Brazil, Peru, Tasmania, New Zealand, New Guinea, and possibly in the near future in Alaska, by allowing transportation by pipeline to the coast. It also obviates the need for enormous port constructions since slurry can be loaded at deep-water off-shore moorings by pipelines. The use of pipelines for the discharge of slurry may, in addition, provide the solution for some steel mills within a short distance of the coast which are currently suffering from higher transport costs on ore compared with coastally sited plants. In 1968 slurry was already affecting locational criteria in Europe, as instanced in the report that Dutch steel interests were planning to site a steel plant on the gasfields of northern Holland which would be supplied by piped ore slurry discharged at North Sea moorings.[11]

Finally, one might recall that the increased size of bulk carriers has not only enabled Australia, once geographically remote, to become a major supplier of ore to Europe but the sheer size and weight of these vessels has changed the geographical assessment of the Arctic as a resource base, for these ships when fitted with ice-bows could break through Arctic ice to reach the vast mineral deposits of that region at any time of the year.

Unitisation

The objective of unitisation is the carriage of large quantities of heterogeneous goods from origin to destination without delays on transferring between modes, without inspection or other administrative encumbrances on crossing international boundaries, and without break of bulk until arrival at, or close to, the final destination. The last characteristic is particularly important for the maritime sector, as it was the handling and stowage of individual items of cargo which kept the productivity of cargo liners at a low level compared with other ships. The manual handling of small single items also rendered cargo vulnerable to breakage and pilferage, thereby raising surveillance and insurance costs.

By packing the goods in a standard container protection is afforded, and mechanical handling and rapid intermodal transfers are made possible. For these reasons the container conforming to ISO dimensions 3·048 m (10 ft), 6·096 m (20 ft), 9·144 m (30 ft) or 12·192 m

(40 ft) in length by 2·438 m (8 ft) in width and depth is currently the most widely adopted unit. The majority of containers are simply steel-framed and aluminium or steel-clad boxes fitted with doors, but for special trades there are refrigerated, humidity-controlled, open-topped, and several other types, including a collapsible variety used in trades where non-containerisable cargoes are carried as back-loads.

The pallet constitutes another means of unitisation. This is merely a portable wooden platform on to which goods are stacked and strapped. It is constructed so as to facilitate movement by fork-lift trucks. An even simpler unit load is that of pre-slung cargo, whereby goods are made up in slings and are stored and carried without removal of the slings until the final destination is reached. Both palletisation and pre-slinging were in use before containers. The most recent development in this field is the barge carried on board ship. The cargo is stowed in several barges either as single items, pre-slung, palletised or, less frequently, containerised; and the barges are lifted on and off special vessels by shipboard crane or elevators.

Types of unitised vessels

The first purpose-built ocean container ships were of the cellular type in which about two-thirds of the containers are stacked in cellular compartments and one-third is carried on deck. Those built between 1966 and 1969 for the British Consortium Overseas Containers Ltd (OCL) were designed to accommodate about 1,300 containers arranged in nine rows across the ship and nine tiers high (six below deck and three above). They were, like most other container vessels, built without container-handling gear; this was installed in the form of gantry cranes at a few ports. The first OCL container vessels had speeds of 21 knots and sharply flared bows to cast seas clear of deck-loads of containers; they were soon modified to carry 1,510 containers.

Since 1969 cellular container ships have increased in size and speed. The second-generation OCL vessels carry at least 2,300 × 6·096 m (20 ft) units (only 352 above deck) and have speeds of 26 knots. Their size and dimensions have been dictated by the limits of the Panama Canal, which allows a maximum overall length of 289·55 m (950 ft) and a beam of 32·30 m (106 ft). The maximum design draught of these bigger vessels is 13 m (42 ft 7 in).[12] Fig. 12 provides a basis for comparison between container ships and other large vessels.

For ships operating through the Panama Canal the above dimensions represent maximum size and their deep draughts limit the number of ports to which they may operate. The vertical cellular design of these ships and consequent wide hatches means that they lack the longitudinal strength provided by continuous decks and are thus subject to longitudinal stresses and torsion, so that even on routes which do not involve canal transits there may be physical limits

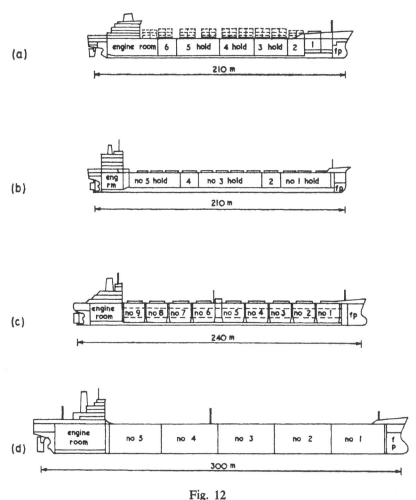

Fig. 12
Ship types, sizes 200–300 m. (a) Container ship (*Encounter Bay*) 21,750 dwt, (b) Ore carrier (*Nuolja*) 72,500 dwt, (c) OBO (*Varenna*) 101,600 dwt, (d) VLCC (*King Haakon VII*) 220,050 dwt.

on the size to which cellular ships may be built. The size of cellular container vessels may also be limited by the increased crane cycle necessary when working deeper and wider vessels, but one must not underestimate the ingenuity of crane manufacturers. A cellular ship with a capacity of 3,000 containers, for example, will have an average crane cycle of three and a half minutes. The time taken to discharge and load such a ship would be fourteen and a half crane days; if four cranes were used, port turnround time would be three and a half days compared with two days for the 2,000-capacity vessel. In order to overcome these particular constraints on size several designs of high-capacity non-cellular container ships were being considered in 1971.[13]

Roll-on/roll-off vessels provide alternative means of carrying containers. One difference between the two types of vessels relates to space utilisation. The lift-on/lift-off cellular ship has about 20% loss of cargo capacity due to the space occupied by cell structures and the containers themselves. The roll-on/roll-off vessel has 30 to 40% loss of capacity as containers are stowed on trailers the below-axle area of which is usually wasted. The roll-on/roll-off ships do have certain compensating advantages, for they are fitted with stern openings and have ramps over which cargo is conveyed and this renders them very flexible vessels able to trade to places which are without special lifting equipment. Originally they were designed for short sea trades where port calls were frequent and transit time at sea minimal. Unfilled space was therefore of less significance compared with the speed of port turnround afforded by this type of vessel. In addition to being able to carry containerised and non-containerised cargoes another advantage of the roll-on/roll-off is the ability to load units weighing up to 100 tons which cannot be lifted in containers. Flexibility in respect of ports and cargo have rendered roll-on/roll-off ships sufficiently attractive to be purchased for ocean-going trades. These vessels are operating on the Australia–Japan service, and on the Australia–Fiji–Honolulu–North America route.

Several unitised ships incorporate lift-on/lift-off and roll-on/roll-off characteristics. The 18,500 dwt vessels of the Atlantic Container Lines (ACL) have cellular holds and deck space for a total of 750 containers, plus intermediate decks for 1,000 cars and the carriage of heavy equipment and loads of up to 300 tons which can be rolled on and off over stern ramps.[14] The facility to work containers by shore gantry cranes and simultaneously load and unload by roll-on/roll-off methods speeds turnround. In this combined system there must be considerable space on shore for the stacking of containers and the marshalling and circulation of vehicles.

Another ship which is increasingly found on relatively short distance voyages is the pallet carrier. This has been in service for some years in the trade from Scandinavia to Britain and between Britain and the Canary Islands. It has been adopted in other parts of the world where small loads are handled at frequent intervals. The pallet ships have stern and side openings and the pallets are brought to the load deck by elevators where they are handled by fork-lift trucks on board ship and on the quay. The loss of cubic capacity is only about 5% for cargo stowed on pallets, but warehouses, and sometimes covered quay areas, are required for efficient working in bad weather.

In contrast to the pallet ship, which can be a converted conventional vessel, the barge-carrying vessel (BCV) is a completely new concept involving very big units. The two types of barge carriers for which there was a demand in 1971 are the 'Lash' (lighter aboard ship) and the 'Seabee'. The first of the Lash ships, *Acadia Forest*, came into service in September 1969. Two further Lash vessels were delivered by mid-1971, and ten others were then being built. In the Lash vessel the barges are stowed one on top of the other in athwartship cells and are lifted over the stern by a 510-ton shipboard traversing gantry crane. The *Acadia Forest* carries seventy-three barges, each with a capacity of 370 tons. They are loaded and discharged at a theoretical rate of one every fifteen minutes, though in practice about thirty minutes is required. The second Lash vessel to be delivered, the *Lash Italia*, is equipped with a container crane in addition to the barge gantries; both of these cargo-lifting mechanisms can operate simultaneously and achieve a cargo rate of up to 1,800 tons per hour in barge and container loads. The BCV is independent of quays and port equipment, it requires only buoy moorings, sheltered water, and tugs. The barges are towed from ship to discharging area and there the cargo is unloaded by cranes. With loaded draughts of only 2·74 m (8 ft 9¼ ins) in fresh water deep penetration of inland waterways and rivers can be achieved.

The 'Seabee' type of BCV, the first of which was launched in mid-1971, carries thirty-eight barges each with 850 tons of cargo. These are handled two at a time, by a 2,000-ton stern elevator on to which the barges are manœuvred. Both Lash and BCV can be converted to full container ships or to bulk carriers when necessary.[15]

Economics of unit loads
The economic gains from unitisation have to be assessed in terms of reduced total distribution costs for shippers as well as reduced operating costs for shipowners. By facilitating transfers between road, rail, port terminals and ships, the time of goods in transit is cut down. Capital turnover is thereby speeded and inventory costs reduced; also, by providing protection for goods, packaging costs are minimised. These are obvious gains to shippers; less tangible advantages lie in the benefits to be derived from a greater measure of certainty regarding the arrival times of goods at final destinations. For the shipowners the capital investment in container ships and barge carriers is very considerable. However, fast turnround means a

reduction in the number of ships required to deliver a given quantity of goods. Similarly, in spite of the high capital costs of container berths port handling charges per ton of cargo are reduced with increased throughput.

It is possible to attach figures to some of the above characteristics and to compare unitised with conventional vessels, but it is not as yet possible to make many firm comparisons between the unit load systems. Capital outlay in the container ship sector has been very substantial and building costs have accelerated. The six 27,000 g.r.t. cellular ships built for the Australia–Europe trade cost £30 million in 1967–9. Containers cost about £1,000 each and every vessel requires on average two and a half sets, for use simultaneously at sea, in port and in the hinterland. Arthur D. Little Ltd estimated an investment in the maritime sector of containerisation of about £700 million between 1966 and 1970.[16] This was probably an underestimate, since twenty-five container vessels, each of 2,000 capacity, ordered for the Europe–Far East trade in 1971 cost £310 million, including containers.

Annual capital costs on investment are thus greater per deadweight ton for container ships than they are for conventional vessels; but, depending on load factors, capital costs per ton of cargo carried per annum are lower in the container system. Many of the operating costs of container ships do not differ substantially from those of automated bulk carriers of a comparable size and speed, and certainly crew costs on container vessels are comparatively less than on conventional ships. Insurance premiums may be higher than those for conventional vessels, since the container ship spends more time at sea, and is thereby exposed more frequently to hazards. There may, in addition, be insurance charges relating to containers carried on deck. The average value of the contents of a container on the North Atlantic is £5,000 and the container itself is worth £1,000. Several have, in fact, been lost overboard in the Atlantic, although this would be an unlikely occurrence when carried on a purpose-built container vessel.

One major difference in operating costs between the container ship and the conventional cargo liner lies in the greater proportion of fixed costs on the former. Whereas cargo-handling costs are usually considered as variable items in conventional shipping, this is not so with the container vessel. The loading of containers must proceed, and the same time is involved, whether the containers are full or empty. This, as will be seen below, has an important bearing on the nature of competition in the container-ship trades.

In discussing the operating costs of various ships frequent reference was made to economies of scale. These do function in containerisation, but there are more constraints on the size of container vessels than there are on oil tankers and bulk carriers. The structural problems and the limitations imposed by the crane cycle on increased size of ship has been referred to. There are also difficulties in port areas due to a shortage of space for stacking and marshalling large numbers of containers arriving at the same time. The commercial constraints may be even more stringent for warehousing and inventory costs incurred in receiving very large shipments of general cargo may more than cancel out gains from economies in scale. In 1971 a Japanese consortium had plans for a 30-knot 3,000-capacity container ship, but few companies appeared to be contemplating sizes beyond 2,500.

In contrast to the container vessel, the pallet ship can avoid some port costs when there is little cargo available, for empty pallets can be loaded relatively quickly and on some trades expendable pallets are used. Productivity per gang hour is about 60 tons on pallet ships; while this is less than the 200 tons average on the container vessel, it is a very marked improvement over the 15 tons typical for the cargo liner. Fast ship turnround, and reductions in the amount of intermediate handling of small quantities of goods as they move from origin to destination, provide the basic economies in palletisation.

When it comes to the BCV very high initial investment is required. The *Acadia Forest*, for example, with two and a half sets of barges cost $17·5 million. The economies lie in the fact that the very expensive mother ship can be kept moving between ports while the low-cost barges are unloaded at berths in waterway hinterlands. Costs of port facilities and warehouses are reduced by this system and many of the problems of port congestion eliminated.

It is difficult to make comparisons between unitised systems because each is suited to a particular task or to a particular economic and physical environment. As a basis for comparisons UNCTAD has made various assumptions relating to size of ship, distance, number of ports, and so on, to produce a model of capital costs, operational costs, and cargo-handling costs for different types of unitised systems. While admitting that different assumptions would produce somewhat different results the UNCTAD report nevertheless emphasises that the figures used are based on actual data, the orders of magnitude in the comparisons are therefore probably realistic. In summary form, costs based on trade from door of developed country with highest costs to quay of developing country with lowest costs in dollars per freight ton of one cubic metre were as follows: conventional ship $23·11, pallet $12·78, Lash $11·49, container $10·87.[17]

Ownership of unitised vessels

Before 1966 container vessels, often converted tankers, were operating on the US–coastal and US–Hawaiian services, and roll-on/roll-off ships were employed in the short sea trades of north-west Europe. In 1966 several countries engaged in the north Atlantic trade adopted containerisation, and in 1968 container vessels were introduced to the short sea trades of Europe and to the North America–Japan route. In 1969 container services commenced between Europe and Australia, and ships were being built for the Europe–Far East services. The Lash ships came into operation in 1969 and 1971 on the United States–Europe and US–Mediterranean routes. Development of unitisation was thus rapid. By mid-1966 there were seventy-three part-container, full-container and roll-on/roll-off vessels (about 1,000 dwt) totalling 900,000 dwt; by the end of 1970 there were 475 such vessels aggregating about 4·8 million dwt. Six barge-carrying vessels aggregating 161,000 dwt were operating and there were twenty-seven BCVs on order in that year. Most of these unitised vessels were owned by a few economically advanced countries.[18] Those owning more than 3% of world tonnage in 1970 are shown in Table 19.

Table 19

OWNERSHIP OF UNITISED VESSELS 1970
(% share)

	Ships less than 5,000 dwt	Ships above 5,000 dwt	BCV
Australia	4·3	3·6	
Italy	3·4	—	
Japan	—	5·4	
Netherlands	3·6	3·2	
Norway	—	4·5	53·4
Sweden	3·4	4·6	
UK	15·8	11·5	
USA	—	51·6	46·6
W. Germany	48·2	5·4	

Source: *Container Ship Register 1969–1970* (AS Shipping Consultants, Norway 1971).

The enormous capital expenditure necessary for container and barge transport has excluded many countries from entering this field. In most of the countries which have adopted unitisation it has been necessary to form consortia in order to concentrate capital and to regulate competition. In the trade between Europe and Australia, for example, the OCL group is made up of four British shipping lines, and ACT (Associated Container Transport Ltd) comprises five British lines. On the north Atlantic run ACL is a consortia of British, Dutch, French and Swedish shipping interests.

The many Japanese shipping companies in existence in the 1950s formed six consolidated groups several years ago, and the six appear to be concentrating more closely into two consortia as container-ship operators. In the Far East–Europe trade multinational consortia have been established representing Japanese, European and British lines. Only US shipping companies, some of which are in receipt of subsidies, have tended to act independently rather than in consortia, although they cooperate with one another as conference members.

The speed of development in the container-ship market has brought with it the danger of overtonnaging and the possibility of freight-rate wars. The likelihood of rate cutting in an overtonnaged trade is intensified by the structure and cost characteristics of the container system. It will be recalled that in the conventional liner sector conferences effectively regulated freight charges, while competition between companies for cargoes remained possible in terms of speed of delivery, quality of service and care of cargo. In the container market there are few real differences between services on which shippers can base a choice. This tends to focus interest on freight rates as a basis of competition, and as there is little to be saved from operating with empty containers the temptation to reduce freight rates in order to attract more cargo is reinforced.

On the Europe–North America trade overtonnaging and rate cutting has occurred. In 1970 the container capacity on this route was 4·3 million tons each way whereas total containerised cargo was a maximum of 3 million tons in any direction. Additional new vessels with higher speeds threatened to raise overtonnaging to more than 50% in 1971. Because of the inability of the conference to maintain rates the ACL group withdrew from three north Atlantic conferences in 1970. In the same year the American Moore-McCormack Line withdrew its vessels from the Atlantic trade, and the US Matson Line did likewise in the Pacific for the same reasons.[19]

The north Atlantic route has been particularly prone to overtonnaging for the US Federal Maritime Commission has, until recently, been opposed to conference restrictions on entry and at the same time the US anti-cartel laws have prohibited mergers and rate agreements. In 1971 the Atlantic conference members of the seven main transatlantic container lines agreed on rate maintenance and the sharing of traffic, the FMC appeared to sanction these moves, and as a result ACL reapplied for conference membership.[20] There was also a trend at this time towards fewer but bigger conferences covering greater areas. A new transatlantic conference agreement, for example, covered the US east coast–northern Europe–UK and Mediterranean; and on the Europe–Australia trade three conferences were replaced by one. These developments in the ownership and operating structure of unitised shipping indicate that strong consortia linked in conference will maintain rates, and they confirm the Rochdale Committee's view that the operators of unitised services in the future, though in the position of being able to provide a greatly improved service to shippers, would seem inevitably to have greater potential monopoly powers than are now available to conventional liner operators organised into conferences.

Impact of unitisation

It is too early to state with any degree of certainty how unitisation will affect the distribution and nature of economic activities on a world scale, but certain trends are discernible and these are briefly considered.

The number of ships moving on ocean routes will certainly be reduced. On the Australian trade conventional cargo liners spend 50% of their time in port, container ships spend only 12% of their time in port activities. With this increased productivity nine container vessels are capable of replacing seventy out of the eighty or ninety conventional ships normally employed on the Europe–Australia service. On the other hand, the number of vessels on intra-regional and coastal trades may well increase as the bigger container ships concentrate services on fewer ports. The speed of transfer of the container could also allow coastal services to compete with land transport in some areas.

It is at the ports that most of the effects of unitisation can so far be seen. Increased cargo throughput in less time means fewer, but more specialised, berths, and the economies of scale from railway freightliner distribution also means more extensive hinterlands linked with fewer ports. The nature and extent of port competition in the container age is detailed in Bird (1971).[21]

The impact of unitisation on labour is self-evident. Fewer ships means less seamen employed, but the longer periods of leave to compensate for time spent at sea on unitised vessels may ameliorate the reduction in total seagoing manpower. On the other hand, most maritime countries would welcome a fall-off in the demand for seamen as labour recruitment has become a major problem. Port labour will be affected by a reduction in the number of men required per unit of cargo throughput, and by a transfer of some cargo-handling activities away from port areas. At a conventional berth, for example, about 100 men can handle 1,200 tons of cargo in seven to eight hours, while at a container terminal twenty men can handle four times this tonnage over the same period. The latter handling rate relates to an intact container, the actual stuffing and unstuffing of the container being carried out at inland distribution centres by non-port labour. Port labour in Britain has been reduced by 40000 per annum between 1971–72.

The through-transport facility of unitisation may affect the flow of cargoes along certain routes. Land bridges by rail across North America and parts of Europe have not as yet materialised as significant links in through transportation. But the transhipment of containers to rail at Montreal for delivery to Toronto and Hamilton has reduced traffic on the St Lawrence Seaway, and containers transhipped at Halifax for rail transport to Great Lakes ports could affect the port of Montreal. In Europe in 1971 the MV *Ivan Chernyck* was transporting containers from Tilbury to Leningrad for onward rail haulage across Russia to Nakhodka and hence by container vessel to Japan. The extent to which this link can compete with container-ship transport direct to the Far East remains to be seen.

The reduction in time of transit and in intermediate handling costs in the through-transport system will no doubt affect the relative competitiveness of various commodities on the world market. In the case of Australia the container was seen as a means of assisting wool to compete with synthetics produced in Europe. The total transport costs of the Australian wool clip was about £75 million per annum in the late 1960s, of which £15 million was attributed to ocean freights and the rest to handling costs and inland transport. The packing of compressed bales of wool into containers close to places of origin was expected to reduce some of these costs, and certainly to prevent them from rising as steeply as in the decade 1958 to 1968 when they increased by 40%.[22]

The BCV system is particularly important in providing through transport for raw material and goods originating from and destined for places on river and waterway systems. As well as generating new cargo from such areas the BCV may revitalise some small ports and help solve the problem of port installation costs in developing countries. Already the sophisticated barge transport system in the US has helped pull industry to riparian sites in the southern states.

Unitisation will undoubtedly have long-term effects on industrial location. The speed of unit transport, which reduces economic distance, and the protection which unitisation provides for commodities in transit, must increase the degree of freedom in location decisions. The final market, for example, may exert less of a pull on manufacturing industry when transportation time is reduced, as a result labour costs, raw materials, or even climate, may prove more of an attraction.

1. Statistics derived from Fearnley and Egers Chartering Co. Ltd, *Fleet Reviews* and *Trades of World Bulk Carriers* (1966–70).
2. Ouren, T., and Sømme, A., *Trends in Interwar Trade and Shipping* (Bergen 1949).
3. Benson Report, *Development Coordinating Committee Report* (British Iron and Steel Federation, London 1966).
4. Westinform, *World Shipping Requirements for Iron Ore 1970 and 1975*, Survey 1/68 (August 1968), p. 9, and *The Role of the Combined Carrier in the Bulk Cargo Trades in the early 1970s*, Survey 1/69 (July 1969).
5. TRADA, *Softwood Handling Project*, Final Report (London 1968).
6. Bolton, F. B., 'The Economics of the Bulk Carrier', *Journal Honourable Company of Master Mariners* 32 (December 1969).
7. 'World's Largest Ore Port', *ICHCA Journal* (September 1971), p. 24.
8. Ensor, J., 'Operational Research', *Financial Times* (3 May 1967).
9. Cufley, C. F. H., *World Freights* (British Sulphur Corporation 1968), p. 15.
10. Finniston, H. M., 'Plans for the Future', *Financial Times* (17 June 1968).
11. Cufley, ibid., p. 26.
12. 'Handling of container vessels beyond the 3rd generation', *Europoort* 71/2, pp. 49–50.
13. ibid., 71/2.
14. 'The Atlantic Container Line', *Europoort* 71/1, pp. 19–28.
15. Meeuse, G. C., 'Barge Carriers', *Scheepvaart Zuid* (Holland 1969), and *Shipping World and Shipbuilder* 162 (December 1969), pp. 1713–18.
16. Little, Arthur D., *Containerization on the North Atlantic* (New York 1967).
17. UNCTAD TD/B/C.4/75, *Unitization of Cargo* (UN, New York 1970), pp. 21–30.
18. *Container Ship Register 1969–70* (AS Shipping Consultants, Norway 1971); UNCTAD TD/B/C.4/75 (1970).
19. OECD, *Developments and Problems of Seaborne Container Transport* (Paris 1971), pp. 11–12.
20. 'Battle of the Atlantic truce in sight', *Containerisation International* (July 1971), p. 19.
21. Bird, J., *Seaports and Seaport Terminals* (London 1971).
22. 'Wool prepares for the container revolution', *Financial Times* (16 January 1968), p. 2.

9

COASTAL AND SHORT SEA SHIPPING

All countries with maritime boundaries, or with major river or lake systems, have domestic shipping of some description. The vessels vary from small wooden craft under sail to sophisticated hydrofoils and hovercraft. In some cases they are legally obliged to operate within certain prescribed limits. British home trade vessels, for example, are restricted to voyages between the ports of the United Kingdom, the Irish Republic and the continent of Europe lying between Brest and the River Elbe.

The operational areas of the so-called short sea trades are less easily defined, but in respect of British shipping they include the above-mentioned home trade areas, the ports of western Norway, the Baltic, Iberia, and the western Mediterranean. Short sea ships when trading outside home trade limits are required to employ certificated officers and to maintain higher standards of safety equipment than are currently laid down for home trade ships.

In many parts of the world coastal vessels become short sea ships or even deep sea foreign-going vessels without changes in crew composition or safety equipment. We distinguish between coastal, short sea and deep sea vessels mainly because of their somewhat different functions and because of the different economic environments in which they normally operate.

The role of coastal shipping
Coastal shipping has played an important part in the economies of countries in Europe, America and Asia. In many areas of Europe and North America the importance of this activity has declined in recent years, but in parts of Australia and South America, where there are long coastlines and less well-developed lateral land transport systems,

the role of coastal shipping is becoming increasingly significant with economic development and resource exploitation.

In the United Kingdom the railways have captured most of the bulk commodities that once moved coastwise, and road transport, with its flexible and fast door-to-door services, has taken almost all the general cargo from domestic shipping. Only where ships link a raw material base to industries at seaboard or riverside locations has coastal shipping survived in competition with rail for dry bulk cargoes. Similarly, it is only in services from mainland Britain to Ireland, the Channel Islands, and the Scottish western and northern isles, that domestic ships have increased their activities in the carriage of general as well as bulk cargoes. In the coastal oil trade, by contrast there has been an overall increase in cargo flow, and it is this in particular which is reflected in the rising ton-mile performance of coastal shipping shown in Table 20.

Table 20

UK DOMESTIC FREIGHT TRANSPORT 1952–68
(thou.-mill. ton-miles)

	Road	Rail	Coastal shipping	Pipeline
1952	18·8	22·4	9·0	0·1
1960	30·1	18·7	9·5	0·2
1968	44·0	14·7	15·3*	1·4

Source: after Prest (1970), p. 158.[1]

* Part of the improvement in coastal shipping is attributable to the change in 1967 to calculating the sea performance in statute miles at sea instead of equivalent land milage.

Cargoes of oil now constitute more than half of all shipments in the British coastal trade. These cargoes, which exclude crude-oil transhipments by deep sea vessels, comprise refined products carried by ships of 100 to 1,499 gross tons, although foreign-going tankers of over 16,000 tons also periodically enter the coastal trade. Tankers engaged in coasting supply installations at most of the ports of the British Isles with over 30 million tons of refined products per annum. Pipelines may provide an alternative, and more economic, means of transporting light oil products from refinery locations to markets within about 300 miles, but in the delivery of oils of high viscosity, where pumping is expensive, the use of tankers and barges will continue on the coasts of Europe and the Mediterranean. Chemical and natural-gas carriers are also becoming more common on coastal routes, but the pipeline grid systems may take most of the expanded domestic flow in these commodities.

In the non-liquid coastal trade coal still predominates. Possibly about 70% of dry cargo transported coastwise in Britain is coal, and most of this moves from north-east England and the Humber to the Thames. But British coal consumption declined by over 50% between 1952 and 1968 and this has reduced the size of the coastal dry cargo fleet from about 800,000 to 400,000 g.r.t. Most of the ships are still engaged in the carriage of coal, regularly or sporadically; and in 1971 some of the bigger east coast colliers were making one trip in three to Rotterdam and Amsterdam to load transhipped American coal for Britain, indicating that the level of British coal production has fallen to well below demand. In 1970 there was still about 14 million tons of British coal and coke shipped in the home trade;[2] other dry bulk shipments carried by coastal tramp vessels in 1970 included stone, sand and gravel, slag, grain and feedstuffs, fertilisers, china clay and cement. Taken together, these commodities probably do not exceed 6 million tons.

The coastal tramps, cargo liners and ferries which remain in the coastal trade serve mainly northern Ireland and the northern and western isles of Scotland. Small vessels like the *Loch Ard* (130 g.r.t.), built in 1955, can carry passengers, cars, buses, lorries and 130 head of cattle. New vessels are also multipurpose and provide space for passengers, mail, dry and refrigerated cargoes and containers, as well as for livestock. The range of stowage and handling facilities which have to be provided on these vessels raises their costs, and the seasonal imbalances in passengers and cars reduce earnings. There are thus problems of viability, so that almost all island services are now subsidised, otherwise freight rates would be inordinate. In their evidence to the government for financial support for island shipping, for example, the Shetland authorities quoted manufactured commodity prices 25 to 30% higher in Shetland than in mainland Scotland, because of freight charges. They reported also that the rearing of pigs and poultry in the Shetlands had declined because of the high freight rates on coastal routes.

The government had generally acknowledged the case for support on social grounds for certain shipping services operating to areas separated from the mainland, and in the western isles of Scotland the improved subsidy-supported services have helped develop the tourist resources of islands which were remote or had difficult landing conditions. The days when the captain of a west highland 'puffer' would carry important passengers ashore on his back at low water have thus gone; now drive-on ferries and other modern vessels berth alongside piers in most islands. The extent to which the introduction of a modern vessel can help generate tourist traffic may be illustrated from the Isle of Mull. In 1963 about 73,000 people travelled to and from that island by sea transport. In 1964 a modern ship came into service and 128,000 passengers were carried; by 1968 this had risen to 233,000.[3]

Only a small sector of the British economy depends on coastal services for its spatial linkages; the Norwegian economy on the other hand shows a much greater dependence on this mode of transport. On the long indented coastline of Norway there are many small industrial, agricultural, and fishing communities which are difficult to reach other than by sea. The coastal seaway provides easy access to these settlements by a network of deep-water channels sheltered from the open sea by the island skjaergard, which extends almost the length of the country from Stavanger to the North Cape.

The Norwegian coastal fleet in 1969 ranged from thirteen express passenger ships each of 2,500 g.r.t. and speed 14 knots, to several small modern palletised cargo vessels and a few older wooden craft. There were also many ferries linking roads on either side of fjords, and a number of fast hydrofoils operating over short distances between the principal towns and their adjacent fjord and island areas. In 1965 the total share of coastal shipping was more than 35% of ton-miles in the carriage of goods in Norway.[4]

Problems of the viability of coastal shipping exist in Norway as elsewhere, even where there is less intensive competition from land transport. The costs of operating the ships have been steadily increasing, in spite of low cargo-handling charges in some rural areas where local farmers, or even the crew, work the cargo. The increases in costs are due basically to the high labour component in the operation of most coastal vessels. There is obviously an unfavourable manpower/tonnage ratio on many small ships, unusual hours are often worked involving overtime, and the very high frequency of cargo handling, made necessary by many ports of call during a coastal voyage, raises labour and other port costs. Foss estimates that about 18% of the total costs of cargo ships on the Norwegian coast are attributable to cargo handling (on board ship only, not on the wharf) and 30% to crew costs.[5] These figures are probably about true for most coastal vessels in north-western Europe.

Modern methods of handling cargoes to reduce port costs, and automated engine rooms with bridge controls to minimise crew, have been introduced to coastal services. Owners also try to schedule their vessels to avoid non-working weekends in port, but social factors interpose in this with regard to retaining crews. Also, where passengers are involved schedules have often to take account of their travelling habits. The major Norwegian coastal operator of express

Coastal and short sea shipping 163

services, for example, has had to reconcile the desire of rural travellers for night passages to and from the main towns, which enable them to minimise loss of working time and also save on hotel bills, with the wishes of tourists for day passages. These cannot be easily reconciled by scheduling, so the company simply favours the tourists in the summer and the local people in the winter. Fortunately, the rural communities have relatively good access to the towns by long-distance buses in summer when mountain roads are snow-free.[6]

By contrast with Norway the coastal services of Sweden have diminished very rapidly in recent years. Even the timber trade to Denmark which was carried out by family-operated wooden vessels and small motor coasters has all but gone. Now timber moves by railways which are linked to efficient ferry services across the Sound.

German coastal shipping, while it also suffers from land transport competition, continues to rate as the principal coastal fleet in western Europe. This important role of German motor coasters has been encouraged by the existence of the free port of Hamburg. There cargoes are discharged from foreign-going ships and redistributed to adjacent countries, without customs formalities, via the coast, the waterway and river systems, and the Kiel Canal. German coasters (up to 500 g.r.t.) made up almost 90% of coaster traffic through the Kiel Canal in 1969, and they carried 2·5 million tons of timber to Britain and other west European countries[7]. Many of the German coasters have, along with Dutch vessels, also participated profitably in the British intra-coastal trade, which is open to participation by any nation.

The coal trade on the coasts of north-west Europe, in contrast to that of Britain, has increased since 1960. In this tug and barge transport has played an important role. Belgian owners are employing transport units comprising two tugs with four barges, each of 13,660 tons, to convey coal from Poland to Antwerp through the Kiel Canal. There is in this trade, which was only partly operative in 1970, a problem of obtaining return cargoes for such large units, but grain shipments from France to Poland was seen as a possible solution to the imbalance.[8]

Types of coastal ships

The characteristic British east coast 2,875 dwt flatirons have almost disappeared from the coastal coal trades. These vessels were built to lie low in the water, they had squat superstructures and were fitted with collapsible masts to pass under as many as twelve Thames bridges in order to reach London's gasworks and power stations.

The self-trimming collier and flatiron without cargo gear, and the versatile small cargo liner fitted with cranes, remained until recently the most advanced vessels on the coast. All the ships tended to be grouped into certain sizes, for they were generally built to the upper limits of tonnage classes beyond which it is necessary to carry additional crew as laid down in the paragraphs of official regulations. The predominance of the so-called 'paragraph' ships has resulted in coastal fleets being made up of blocks of vessels in such sizes, 199, 299, 399, 499 . . . 1,599 and 1,999 g.r.t.

While still conforming to the upper limits of the paragraphs coastal shipping has been undergoing various changes in type. Specialisation in construction is now a feature of the coastal trade, particularly in the carriage of naphtha, petro-chemicals, LPG and LNG. Also becoming more common is the coastal container vessel, the pallet ship, and the roll-on/roll-off vessel. These are described under the short sea trade in which they are particularly active (pp. 172–6). But in the coastal services of Britain, including those to northern Ireland and the island areas, very striking advances have also been made in the introduction of modern unitised vessels. Furthermore, as ocean-going container ships begin to concentrate cargoes at a limited number of main ports feeder container services are developing. In the British coastal trade in 1969 roll-on/roll-off services carried 818,000 tons of goods, lift-on unit services carried 1·6 million tons, and conventional vessels carrying containers delivered 389,000 tons.[9] These tonnages must represent a substantial proportion of the general cargo trade moving coastwise and across the Irish Sea.

Unitisation and other technological advances of various forms, from labour saving to time minimising, have greatly improved the efficiency of coastal fleets. According to sources quoted by Øvrebø, in 1948 each deadweight ton of West German coastal shipping transported on average 15·1 tons of goods, in 1954 about 27, in 1960 29·4 and in 1968 no less than 33·3.[10] Some newly built coastal vessels combine advances in cargo handling and stowage to meet particular local conditions. In Australia dual-purpose ships carry containers on one leg of a long coastal voyage and bulk cargoes on the other. On the coast of the United States, where the problem of high labour costs is particularly acute, the direction of technological change has been towards tug and barge systems in the carriage of bulk cargoes. Unmanned barges aggregating 20,000 tons, and tugs employing very small crews, are being used on US coastal waters and river systems. It is likely that tugs and barges will become more common in the carriage of petroleum and minerals on some of the major coastal and short sea trades for dramatic savings on a cost per ton basis appear possible.[11]

Another specialised type of coastal vessel which has made its appearance involves an adaptation of the roll-on/roll-off design for the carriage of abnormally heavy loads. There has been a substantial increase in the construction of heavy indivisible pieces of equipment for use in the chemical, petroleum and power industries. The transportation of these from place of manufacture to destination has presented great problems. In Britain single units of power station plant, weighing more than 300 tons, have been required to be moved from one part of the country to another. Such units exceed the dimensions permitted on the railways, and to move them by road has meant following special routes, the strengthening of bridges on the way, and considerable traffic congestion. The solution has been found in the construction of shallow-draughted ships capable of entering small ports as near as possible to the origins and destinations of the loads and loading and unloading them using a roll-on/roll-off principle and shipboard hydraulic rams as motive power.[12]

Ownership of coastal fleets in Europe

On the coast of Britain tanker shipping is owned mainly by the major oil companies, in the coal trade the majority of ships now belong to one private company, and, to a declining extent, to the electricity and gas corporations. The main company outside the coal trade is Coast Lines Ltd (owned by the P. and O. group) with twenty-nine ships in 1969. In addition there are still several private owners and small companies with one or two vessels; taken as a whole most of Britain's coastal fleet of around 444,000 g.r.t. in 1968 was by 1970 owned or managed by ten companies, including British Rail.[13]

By contrast with Britain, German and Dutch ownership is more diffused. It was reported in 1968 that 1,012 West German ships were owned by 946 companies. Many of these companies consisted simply of a family-unit operating a small ship. Under this arrangement the family sails on the ship as crew, and in some ports works the cargo, thereby enabling low freight rates to be accepted. This obviously is a partial explanation for the survival of greater numbers of German and Dutch coastal and short sea vessels. The trend is for one-ship owners to place their ships under the control of management companies which arrange charters thereby obtaining the benefits of central management and the economies of family operation. In Holland, however, there has been some reduction in the numbers of small conventional vessels in favour of more expensive specialised ships under company ownership. Table 21 provides an approximate division of ownership of small vessels in the European coastal trades by flag, as it was in 1966.

Table 21

MAIN WEST EUROPEAN COASTER FLEETS 1966

	Ships	100–500 g.r.t.
W. Germany	1,001	317·5
Netherlands	903	381·1
Norway	559	146·9
Denmark	444	111·3
Sweden	357	117·4
United Kingdom	171	not available

Source: Øvrebø (1969).

The role of short sea shipping

The area of operation of British short sea shipping extends from Norway to the Mediterranean. There is thus an overlap with home trade limits. Indeed, precise distinctions cannot always be made between short sea and home trade vessels, especially in and around the southern North Sea and English Channel on which this section focuses. This is one of the busiest maritime trading regions in the world. It is crossed each day by hundreds of vessels ranging from coasters to giant tankers, and within a radius of 200 miles from a central position in the southern North Sea lies more than a dozen of the world's major ports and several principal river and waterway systems.

The role of short sea shipping in the economy of Europe is of considerable importance, and for Britain it is vital. Over 41% by value and about 60% by volume of British overseas trade was carried by ships to Europe in 1970.[14] In the seaborne trade between all the countries of Europe about 125 million tons was hauled in 1967, of which between 70 and 80 million tons was transported on vessels below 2,000 g.r.t.[15] The significance of these figures will be appreciated by recalling that the total export trade of the developed world, including the USA, amounted to 480 million tons in 1967.

The physical separation of Britain from the main markets of Europe gives to short sea shipping a special significance in the UK economy. The costs of crossing the narrow sea represent a barrier which has to be overcome in selling exports in the countries of Europe. The report *Through Transport to Europe* points out that a high proportion of British exports could be at risk were there weaknesses in transport arrangements, but also there could be considerable benefits if transport were improved. The report emphasised that continental manufacturers with whom Britain is competing in the extensive markets of Europe do not face the problems involved in sea transport. Exporters from Europe do have to consider costs of sea transport if they wish to reach markets in Britain; but this market is relatively small compared with those of Europe. Access to markets has thus been a major factor influencing British shipping companies to improve sea transport arrangements in the southern North Sea region. For this reason, before considering the technological changes that have taken place in short sea shipping, certain trends in the composition and direction of trade are considered.

Trends in short sea trade

One of the most significant changes in recent years has been the shift in the direction of British trade towards Europe. This is indicated by Table 22.

Table 22
UK EXPORTS BY DESTINATION 1960 AND 1970

	1960 (value £3,789 m) (%)	1970 (value £8,063 m) (%)
EEC	16·1	21·9
EFTA	11·9	15·9
Other W. Europe	2·2	3·6
Total W. Europe	30·2	41·4
Sterling area	38·3	27·5
North America	15·7	15·4
Rest of world	15·8	15·7
	100	100

Source: Based on *UK Economic Progress Report* (14 April 1971), p. 2

It is very significant that the EEC in 1970 took, for the first time, more exports from Britain than did the whole of the Commonwealth. Import patterns indicate similar directional trends. Imports from the sterling area to Britain fell from 34% of the total in 1960 to 27% in 1970, whereas imports from western Europe increased over the same period from 29% to 38%.[16]

These trends are partly indicative of the shift common to most developed market economy countries towards more intra-trade and a reduction in their rate of growth in the demand for raw materials and foodstuffs. In the short sea trades this is seen in the increased flow of high-value manufactured commodities between the industrial countries of north-west Europe, and it is also expressed in the rise of ports on the east coast of Britain to national status for the first time since the medieval period. The growth in exchange of manufactured commodities should not mask the fact that measured by tonnage there is a considerable amount of low-value bulk commodities moving on short sea routes. In 1965 exports of manufactured commodities from Britain probably did not exceed 7 million tons whereas, as Table 23 shows, over 8 million tons of dry bulk cargo

Table 23
EXPORTS AND IMPORTS OF BULK COMMODITIES IN UK–EUROPE TRADE 1965
(mill. tons)

Exports		Imports	
Coal	4·7	Timber	3·8
China clay	1·5	Pulp	3·0
Other materials	0·9	Cement	0·9
Cereals	0·2	Cereals	1·1
Fertilisers	0·8	Fertilisers	2·1
	8·1		10·9

Source: Based on *Short Sea Shipping* (NEDO 1970), pp. 46–7 (including Poland, E. Germany and Iceland).

was exported in that year. The bulk commodities were carried mainly by short sea tramp vessels. Such ships have lost cargoes in recent years due, on the one hand, to the changing nature of market demand and, on the other, to the competition from bigger overseas vessels delivering bulk cargoes from more distant areas. At the same time short sea tramps have gained as a result of the concentration of some of the cargoes of the larger overseas bulk carriers at fewer ports, thus necessitating transhipments.

The most pronounced changes in short sea trade relates to coal and coke. In the pre-1914 era Britain exported 50 million tons of coal to Europe, while now less than 3 million tons is carried from the UK across the North Sea. The demand for coal is actually a good deal higher than the latter figure suggests, for Poland in 1967 shipped 5 million tons of coal to Scandinavia alone, a traditional UK market area for this commodity. The coke trade has suffered in both the supply and demand sides. Coke breeze was, until recently, a common cargo for the 1,599-ton type of short sea trader operating from Britain to Norway. This commodity was produced from the carbonisation of coal at gasworks and was used as a reducing agent in the electro-smelting processes of Norway's fjord industries. New methods of smelting have diminished the demand for coke in Norway, while natural gas has replaced coal at many gasworks and reduced supplies of coke.

It is difficult to make predictions on the future of the coal trade in the North Sea region. The reduction of coal output in Britain from

227 million tons in 1957 to 147 million tons in 1970 was made on the assumption that oil and gas would continue to replace coal. However, the higher oil prices imposed by the crude-oil-producing states, and the rise in the price of United States coal, which has been imported to Europe very regularly in the last decade, may have brought about a revaluation of coal resources in Britain and other parts of Europe. The possibility that coal may be returning to a more competitive position as a source of energy is also reinforced by the use of low-cost barges, and by the much improved loading facilities at places such as Immingham on the Humber which ensures rapid turnround and thus lower transport costs. It does seem likely that the shipowners and the miners were correct in their frequent references to the dangers of allowing coal resources to be reduced with mine closures, and the National Coal Board was wrong in their forward estimate of the demand for coal.

Timber is the cargo which has now assumed an equal place with hard fuels in the short sea trades. Traditionally timber has been loaded at a great number of ports in Norway, Sweden and Finland and carried by small ships across the North Sea. In the early 1960s there were still twenty-two ports in Norway at which timber and timber products were loaded for the United Kingdom, and in Sweden there were about twenty-seven such places. Now packaged timber, and the use of ferry and container services for unitised timber products, have made it easier and cheaper for cargoes to be carried over longer distances by land transport and concentrated at fewer ports for export. As a result in the Swedish timber trade the ships of one major company are loading at three ports instead of twenty-seven, and discharging at Rotterdam, Antwerp and London only instead of many ports in Britain and Europe. These trends are reducing the number of vessels engaged in the carriage of timber, pulp and paper.

The grain trade in the North Sea area is also undergoing changes. Grain exports from Britain, which are reasonably stable, comprise barley for malting and some cargoes of oats. But short sea grain imports to Britain are mainly wheat, which has been transhipped from the ports of Europe, where bulk carriers can discharge quickly into vast storage facilities and often directly on to short sea tramps. These arrangements have enabled 100,000-ton ships to carry grain from Australia to Europe, and for this to be economically transhipped to Britain. In 1968 grain transhipments reached 1·5 million tons. Similar giant grain terminals in Britain will remove the need for transhipments and lead to a further reduction in short sea tramp shipping, but this may also depend on EEC grain policy. As far as the other principal bulk commodities are concerned, such as cement and fertilisers, the trend is towards specialised types of carriers, again reducing employment prospects of the traditional short sea tramp vessel.

The tendency to adopt more specialised types of ships is particularly apparent in the carriage of chemicals. Hydro-carbon feedstocks, potassium chloride, ethylene, liquid gas, liquid sulphur and many additives and solvents are constantly being moved between Britain, Germany, the Netherlands and France. With reductions in tariff barriers there will be increased movement of semi-manufactured chemicals by special tankers and container vessels. It is significant that the chemical industry has twice the growth rate of any manufacturing industry in the UK, and it is expected to generate a high proportion of the trade between Britain and Europe.

As to manufactured goods, the growth in the trade of high-value commodities may be partially indicated by the exchange of motor vehicles between Britain and the countries of western Europe and Scandinavia. In 1960 Britain exported 88,000 assembled and unassembled cars and imported 52,000. In 1968 the figures were 181,000 exported and 71,000 imported.[17] Apart from cars it is extremely difficult to estimate the quantities of various types of manufactured goods transported by short sea shipping, but Table 24 shows

Table 24

CONTAINER AND ROLL-ON TRAFFIC 1969
(mill. tons)

	In	Out	Total
West Germany	0·2	0·1	0·3
Netherlands	1·2	1·1	2·3
Belgium	0·7	1·0	1·7
France	0·7	0·5	1·2
Denmark	0·3	0·1	0·4
Sweden	0·4	0·3	0·7
Other	0·1	0·1	0·2
	3·6	3·2	6·8

Source: *Container and roll-on port statistics, Great Britain* (National Ports Council 1970), p. 11.

tonnages carried by wheeled and container units in 1969. A high proportion of this was general cargo, although some represented transhipments from and to overseas vessels at Rotterdam as a result of the industrial dispute in the port of London during 1969. Also moving across the narrow sea for purposes of transhipment have been heavy indivisible loads carried by special roll-on/roll-off ships. They include high-pressure reaction vessels, used in nuclear power stations and in oil-refinery cracking processes, and heavy turbines, generators and transformers. These are periodically shipped from Britain to Rotterdam on special craft to be loaded on to overseas ships by heavy-lift cranes. Finally, trends in passenger services should be noted. Somewhere in the region of nine million passengers used the shipping services between Britain and Europe and across the Irish Sea in 1969 and they are tending to generate more car transport. About one million accompanied cars were carried by ferry services in 1967 and the rate of expansion was expected to be at over 10% per annum.

Trends in short sea shipping
Up until recently the short sea trade, like the home trade, attracted many one-ship owners. Often a captain-owner sailed his vessel and his family manned it. This was a feature which distinguished this type of shipping from foreign-going enterprises which required greater capital investment in ships and had higher managerial and other overheads. There are still many family-operated short sea tramps, and several small companies still seek charters for two or three ships of less than 2,000 tons.

The ships of less than 2,000 tons operating in the North Sea region are mainly 'paragraph' vessels. Of these there is a high frequency of Dutch vessels around 499 g.r.t. Above this they must comply with certain safety regulations and extra manning, and they fall into a higher class for charges in the Kiel Canal, and this is also the size of ship which can navigate the Rhine as far as Basle. For the same reasons the 499-ton size is a common class of German motor ship. Other paragraph ships range from the 99- to 299-ton Scandinavian vessels to the 1,999-ton Polish ships. A popular paragraph size for UK owners is just below 1,600 tons. The paragraph ships are employed in all the bulk cargo trades, although in the coal trades ships of over 7,000 tons are common. In the carriage of general cargo paragraphs have lost their significance to unitised ships. As unit loads represent the foremost developments in short sea shipping each system is considered in turn.

In 1968 there were about 100 unitised services operating out of thirty UK ports, half of these were on short sea routes to north-west Europe. The division by type of vessel engaged on short sea services in that year was 7% train ferry, 36% roll-on/roll-off, 24% semi-container ships and 30% full container ships.[18]

The train ferries have operated since the 1930s. They are specialised vessels requiring purpose-built berths. The ships have stern doors through which railway wagons are shunted over lines connected between the berths and the train deck of the vessels. Four ships each of 1,100 dwt and three between 1,800 and 2,100 dwt provide a total of up to twenty-four crossings per day on the routes Dover–Dunkirk, Harwich–Dunkirk, Harwich–Zeebrugge. In 1969 they carried 780,000 tons of goods between Britain and Europe. There is a constant imbalance in this trade, with more than two-thirds of the total tonnage of cargo moving on the Europe to UK leg, due mainly to the imports of fruit and vegetables from Italy and Spain which are traditionally transported by rail.

The train ferries continue to provide a valuable service, but the high capital cost of the ships and equipment, the loss of pay-load below the axle level of wagons in the train deck, the high tare weight of the wagons, the inflexible nature of the ships, and the alternative unit services available, make it unlikely that further expansion will take place in the specialised train ferry sector.

Another vessel which has been in operation since the interwar years is the passenger-vehicle ferry. Cars are driven on and off these vessels over bow and stern ramps. Passenger accommodation is provided which must be varied in type to suit day or night services and this, along with the special port facilities required, renders them unsuited to work other than on short sea routes. In these routes their earnings are restricted by the seasonally peaked nature of passenger demand. Several of the ships must, in fact, be laid up during winter and although those on the Mediterranean and north African routes have a longer period of passenger demand, extending into late autumn, they experience more competition from air services than do those on shorter runs.

Closely related to the passenger/vehicle ferry are the roll-on/roll-off ships. They evolved from wartime landing craft and were used initially for the conveyance of loaded lorries and their drivers between Britain and Europe. Now most of the traffic is in the form of loaded trailers, the traction units and drivers frequently remaining at either end of the sea link. British Rail has increased the versatility of its roll-on/roll-off services in order to cope with the seasonal fluctuations in passenger traffic. In 1969 BR introduced the ferry *Vortigern* which could carry cars or railway wagons. This ship now operates a car/passenger service between Dover and Boulogne in summer, and a rail-freight and sleeping-car service between Dover and Dunkirk in winter. The capital cost of the ship was £2·5 million and, like most vehicle/passenger ferries, it requires a crew of between 60 and 70. Fig. 13 shows that in size this new ferry approximated to that of ocean-going vessels.

Statistics are not available which isolate the short sea trade from total trade. But out of the total 'wheeled traffic' recorded in British trade for 1969 about 5 million tons of goods were handled at twelve

Coastal and short sea shipping

British east and south coast ports—much of this belonging to the short sea sector. These goods were carried by about 500,000 wheeled units, which included controlled temperature vans and tanker trailers.[19]

The main competitor to the roll-on/roll-off vessel is the cellular container ship. The two principal British Rail container ships on the Harwich–Zeebrugge and Harwich–Rotterdam services can each

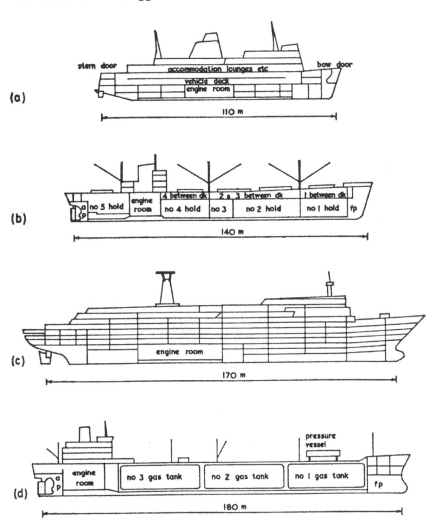

Fig. 13
Ship types, sizes 100–200 m. (a) British Rail, train, car and passenger ferry (*Vortigern*) 4,371 g.r.t., (b) Tramp ship (SD14) 14,200 dwt, (c) Passenger liner (*Hamburg*) 24,950 g.r.t., (d) Liquefied gas carrier (*Faraday*) 25,149 dwt (31,210 m³).

carry 140 ISO containers 9·144 m (30 ft) in length, or a combination of 6·096 m (20 ft) and 12·192 m (40 ft) units. Their schedules provide for a crossing time of seven hours, and for five hours in port at either end.[20] In 1969 cellular container ships operating from Harwich carried about 800,000 tons of goods. Because they require expensive port equipment, and extensive back-up areas for stacking containers, such ships will tend to concentrate on high-density routes between few ports.

By contrast with the cellular container ships the numerous pallet carriers which are operating across the North Sea are very flexible in respect of ports of call. Consignments of from 1 to 3 tons are stacked on pallets and loaded and discharged through the ship's side doors with the use of fork-lift trucks. On some ships cargo is also simultaneously handled through stern openings and through conventional hatches; on the larger pallet vessels there are automatic elevators and cargo conveyor belts running from bow opening to stern opening. With such ships cargo handling rates of 200 to 300 tons per hour are achieved without the cost of special port facilities—although in some older ports which have berths encumbered by railways lines, and very narrow quays, discharging and loading from side ports is slowed down considerably.

Finally, on the shortest routes across the English Channel hovercraft are proving viable competitors to ferry services. The SRN 4 is capable of ten return passages per day between Dover and Boulogne carrying thirty cars and 250 passengers. It is still too early to speculate on the extent to which hovercraft services will successfully expand into cargo haulage, but with the anticipated increases in demand for car and passenger transport between Britain and Europe both British Rail and Hover-Lloyd are planning extensions to their car and passenger hovercraft services.

Comparison of the systems
The various unitised services on the short sea trades should be capable of functioning within the distributional system in such a way that loads of different sizes, and with different landward modes of collection and delivery, can be transported efficiently at lowest total cost.

For very small consignments of less than 2 tons, and those destined for several places, the most appropriate system would appear to be palletised ships linked to road-transport vehicles. This system has proved particularly successful in the Scandinavian–UK trade which has a relatively high frequency but low-density flow of goods.

On the much denser southern North Sea routes the roll-on/roll-off ships connect the road systems of Britain and Europe. They make possible a fast link from door to door without intermediate transfers of the goods at collecting or distribution points, or between road and rail, although depots are used to make up full container loads in some areas. In general roll-on/roll-off traffic also allows rapid clearance from ports and, when necessary, deliveries direct to premises in urban areas, without mechanical handling equipment being involved. They can thus provide the most rapid form of delivery from origin to destination and are particularly suited for the carriage of perishables. The disadvantages of the roll-on/roll-off system are the high cost of the ships, loss of payload in the stowage of trailers, and the problems of queuing and poor road access to some ports due to traffic congestion. However, because the cost of road haulage tends, in Britain, to rise steeply for distances of over 150 miles there is a tendency for road-oriented roll-on/roll-off shipping to operate from and to a wide spread of ports, each with regional hinterlands, and this helps to relieve congestion as well as to encourage local export industries.

The cellular container ships are more rail-oriented than the roll-on/roll-off. As their port facilities are costly, over £1 million per berth, they must therefore have a high throughput in order to spread capital costs. This means concentration of cargoes drawn from an extensive national and multi-national hinterland. For long-distance haulage of large quantities of goods the economic characteristic of the railways, which have diminishing costs per ton-mile over distances exceeding 200 miles, are utilised. Freightliner trains hauling ninety or so containers link with container ports in Britain and thus maintain the required throughput.

The cellular ships have the advantage of less broken stowage than the roll-on/roll-off trailer services, but they are often faced with the problem of grouping various items of goods into full container loads and the need to employ road services for feeder purposes. Grouping involves handling commodities at inland clearance depots (ICDS) as well as at the places of origin and destination. On the short sea route the cellular system is best suited to the haulage of full loads from origin to destination, as in the transport of unassembled cars from factories in the midlands of England to factories in Belgium, a service which by using company freightliner trains and cellular ships can achieve delivery within twenty-four hours.

The various systems have undoubtedly raised the productivity of sea transport tonnage on the short sea routes. This has also meant overtonnaging as old and new ships have competed for freights, and further advances in technology have been achieved. When, for example, British Rail introduced their cellular ships in 1968 it was thought that this spelt the end of many roll-on/roll-off services. One or two operators have indeed gone out of business since then, but most have survived under conditions of severe freight-rate reductions. In 1967 the freight charged on a 12·192 m (40 ft) container was £70 in the Felixstowe and Harwich to Rotterdam routes, it was £50 in January 1968, and £40 in June 1968.[21]

Because of freight-rate cutting it is difficult to make cost comparisons between the systems. Each can carry out certain transport tasks at least marginally better than the other, but there is no way of allocating traffic between them to achieve rationality of use other than through shippers' choice, which is made on the basis of freight charges and performance. At times of overtonnaging freight charges per container are not necessarily closely related to costs, and shippers' perception of the value of various through-transport services is not as yet perfectly clear. However, the claims that British Rail has been cross-subsidising their shipping sector, and giving preferential treatment in rail haulage to goods destined for cellular ships, has, according to several authoritative bodies, been without foundation—although some private roll-on/roll-off operators have maintained that this is the case.

Future developments in short sea trade

Several possible developments in short sea vessels have been alluded to above but some further comment is made on trade and its likely distribution between the transport modes.

The reductions in freight rates as a result of intensive competition between operators has so far been favourable to exporters. It will no doubt remain so, and provided it does not result, in the long run, in an oligopoly of a few successful survivors then the low rates will continue to generate increased traffic. The tendency in the short sea unitised services to charge f.a.k. (freight all kinds) for container loads, rather than *ad valorem* rates for commodities, may have both a beneficial and retrogressive effect in terms of cargo flow. Shippers of high-value goods in containers will clearly gain, those exporting commodities of low value will lose. The latter effect might result in a reduction in trade of low-value products, or possibly more use of slower forms of transport such as barges.

Britain's entry into the Common Market is an obvious factor affecting the future volume of trade. The EEC has not as yet a definite policy on shipping. There is multilateralism in respect of the carriage of cargoes from and to overseas, and British shipping has a fair share of these. If the policy of free competition between ships and ports is advanced, then, with growth in trade, and increased congestion in the Strait of Dover, there is the possibility of transhipments being made to short sea vessels at British east coast ports from foreign-going ships unloading at advantageous focal ports on the west coast of Britain, using land-bridge rail and road linkages between them. There may also be increased movement of part-assembled equipment and semi-processed materials between member countries in the EEC as differential growth takes place at various industrial locations to take advantage of differences in factor endowments.

Finally, for the future there is the projected Channel tunnel and the bridge between Denmark and Sweden to consider. The first has been seen as a partial solution to the build up of road vehicles requiring fast transit on the cross-Channel routes. On the other hand the concentration of traffic at rail terminals at either end of such a tunnel could defeat its purpose. The alternative of more vehicle ferries does allow the spread of traffic over several ports thus relieving road congestion, but at the expense of greater density of shipping in the already congested English Channel and southern North Sea. The effects of a Channel tunnel on short sea shipping remain in the realms of speculation at this point in time. The second projected land transport link in northern Europe is by a road or rail bridge between Denmark and Sweden. This could affect short sea traffic trading to Scandinavia. With reduced land transport costs it could prove more economic to land goods at the nearest continental port and forward them by freightliner to Sweden. These developments emphasise the trend towards long land, short sea linkages.

1. Prest, A. R., (Ed.) *The UK Economy* (London 1970).
2. Stephenson Clarke Shipping Co., Newcastle upon Tyne (correspondence 1971), and *North of England Shipowners' Association Annual Reports* (1957–70).
3. Highland and Islands Development Board, *A Ferry for Orkney* (1969), local press and Skewis, W. I., *Transport in the Highlands and Islands* (Glasgow 1962).
4. Foss, B., 'A Cost Model for Coastal Shipping', *Journal of Transport Economics and Policy* (May 1969), pp. 195–6.
5. ibid., pp. 206–8.
6. Lavik, J., *Services and Schedules*, Paper at UN Interregional Seminar on coastal shipping (Solstrand, Norway September 1969).
7. Dipl-Volkawirt-Schnuis, 'Coastal Shipping and the Kiel Canal', *Nord-Ostsee Kanal* (Kiel 1970).
8. Dipl-Volkawirt-Murl, 'Polish Sea Traiffic in the Eyes of the Kiel Canal', *Nord-Ostsee Kanal* 3/4 (Kiel 1970), pp. 18–19.
9. *Container and roll-on port statistics, Great Brtain* (National Ports Council Part i, 1970), p. 18.
10. Øvrebø, Stein. H., *Short Sea and Coastal Tramp Shipping in Europe* (Institute of Shipping Research, Bergen 1969).
11. *A Comparison study between Tug/Barge Systems and Conventional Ships in selected UK–Continental Trades* (National Ports Council June 1970).
12. Farrall, R., et al., 'Land and Sea Transportation of very heavy power-station equipment', Paper, Institute Electrical Engineers (April 1968).

G

13. Rochdale Report. *Committee of Inquiry into Shipping* (HMSO, Cmnd. 4337, London 1969), para. 239.

14. 'Britain's Overseas Trade in 1970', *UK Economic Progress Report* (HMSO 14 April 1971), p. 2.

15. Øvrebø, H. Stein, *Short Sea and Coastal Tramp Shipping in Europe* (Institute of Shipping Research, Bergen 1969), p. 27, and *Through Transport to Europe*, Economic Development Committee for the Movement of Exports (HMSO, London 1966).

16. *UK Economic Progress Report*, op. cit.

17. *Short Sea Shipping*, Economic Development Committee for the Movement of Exports (HMSO, London 1970).

18. ibid.

19. *Container and roll-on port statistics*, op. cit.

20. Harrington, J. L., *Short Sea Container Shipping* (Royal Institution of Naval Architects, 10 November 1969).

21. UNCTAD TD/B/C. 4/75. *Unitization of Cargo* (UN, New York 1970), p. 99.

10

SHIPPING AND THE DEVELOPING COUNTRIES

The developing countries include most of the states of Africa, the Middle East, Far East, south-east Asia, Latin America and the majority of island groups in the Pacific. Their common characteristics include a dependence on a small number of primary products for sale on the world markets, a high percentage of population engaged in cash and subsistence agriculture, low levels of saving and investment, and slow rates of growth, or in a few cases even decline, in average per capita income in the face of rising populations.

The dependence of the developing countries on a limited range of exports renders their economies vulnerable to fluctuations in world commodity prices. Annual variations of more than 10% are common for several commodities and the free market price for sugar, to quote an extreme case, has varied from £12·25 to £100 since 1945. Temporary depressions in the purchasing prices of foodstuffs and raw materials come about through oversupply and from reductions in demand arising from economic recession in the industrial countries. Temporary peaks in prices arise from shortfalls in the output of major producers due to adverse weather or strikes, or to stockpiling in industrial countries for military purposes. These fluctuations are superimposed on a long-run deterioration in net barter terms of trade of developing nations.[1]

In the last decade or so technological progress in manufacturing processes has meant less raw material input per unit of output; also, the cost of production of synthetics has been reduced as a result of research and of economies of scale in industry. Synthetics have tended to set a price ceiling for cotton, wool, flax, jute, silk, hemp, manila, rubber, copra and vegetable oils. This is particularly so now that many industrial users of raw materials have acquired interests in synthetic manufacturing plants. Furthermore, the rising wage-price spirals in the industrial countries have contributed to the increased costs of manufactured goods imported by the developing nations. These processes of deterioration in the terms of trade of some regions of the developing world have been identified for over half a century. A. K. Ho points out, in relation to south-east Asia, that a given quantity of primary products would pay, on the eve of the Second World War, for only 60% of the quantity of manufactured goods which it could buy at the beginning of the nineteenth century.[2]

The overall deterioration in the purchasing power of exports, in terms of imports, appears to have been halted in recent years; although sudden price changes could again alter the world balance. In these circumstances it is natural that all the developing nations are taking measures to expand and diversify their exports and reduce their imports. Many of them see in shipping a way of achieving both aims. A shipping industry can earn overseas currency by providing services to other nations, and it can conserve overseas payments by substituting national flag for foreign flag carriers in the import and export trades.

The extent to which the services of foreign tonnage represent a drain on the balance of payments of the developing countries was quoted by Richards as $1,800 million in 1961;[3] others put it at a much higher level.[4] The decision to adopt national fleets as a solution to balance of payments problems is reinforced by the contradiction, which developing countries see, between the origins of cargo and the ownership of shipping in the world. Between them the developing nations generated 41% of world seaborne trade in 1969 but owned a mere 7·6% of world shipping. Of the new tonnage on order in that year 58% was for developed countries, 25% for flag of convenience owners and only 5·2% for developing countries.[5] It must be remembered, however, that many of the latter were in the market for second-hand vessels.

In spite of the comparisons and contrast frequently made between a country's trade and the size of its merchant fleet there is no necessary connection, *ipso facto*, between the trade of a nation and the ownership of the vessels conducting the trade. Nor does it follow that foreign currency will be conserved by operating national flag ships in place of foreign vessels. Compared with operating a national fleet more significant gains may, for example, be made by obtaining freight-rate reductions from foreign shipping companies; or by changing the form or type of cargo or by improving national ports. Before discussing the move towards national fleets it may be useful, therefore, to consider in more detail how freight rates affect the economies of the developing countries.

Freight rates and exports

One of the most important questions in relation to the carriage of a country's exports is, who pays the freight charges? This question is not amenable to a straightforward answer, for the party paying the freight does not necessarily bear the cost. Much depends on the elasticity of demand for the commodities. If the demand for a commodity is highly elastic, that is to say, if a given percentage price increase results in a more than commensurate percentage fall-off in demand, it is likely that the sellers will have to bear most of the cost of any increases in transport charges. This is often the case with the exports of less advanced nations.

One of the reasons for the elastic character of demand for the exports of the developing countries is the existence of artificial substitutes. Man-made synthetics were resorted to on a major scale during the Second World War when many traditional sources of natural fibres, rubber and vegetable oils were rendered inaccessible to the industrial nations. Since then there has been a phenomenal rise in the production of synthetic material such as polypropylene which is replacing many natural fibres, and polyisoprene which competes with natural rubber.

The man-made substitutes are generally produced relatively close to the industries which use them. Natural products by contrast have often to overcome the time and costs of a sea passage before reaching markets. The exporter in the developing country thus finds that competing synthetics frequently have easier access to users, and synthetics can be adjusted more readily in quality and quantity to meet the demands of industry—hence the prime significance of efficient and cheap sea transport for the trade of the developing countries.

Because synthetics put a price ceiling on many raw materials any increases in freight rates tend to be passed back to the sellers rather than on to the buyers. It follows that, in the view of exporters in developing countries, the importers in the developed areas are relatively unconcerned about rises in freight rates for raw materials. Thus in many instances commodities are bought f.o.b., that is the buyer pays to the seller the market price ruling in the industrial country less the freight rates, or a proportion of these. The buyer thus purchases the material on f.o.b. terms and ostensibly pays the cost of carriage, but without bearing the burden of fluctuations in freight rates. As an f.o.b. purchaser he has also the right to nominate the ship in which to carry his goods. In the view of sellers in the developing countries not only can this system of nominating ships constitute a subtle form of flag preference, but, as the buyer does not normally have to bear more than a little of the freight costs, he is less diligent than he might otherwise be in securing the lowest possible freight charges.

There is no doubt that freight rates have been rising. The German liner freight index shows an increase of 80% between 1960 and 1968.[6] For the developing countries this means that when the demand for raw materials is favourable, and the market prices are correspondingly high, freight rates reduce returns to the producers by only a small percentage. But when prices are low freight charges assume a significant proportion of the value of the products. Table 25 shows

Table 25

LINER FREIGHT RATES AS PERCENTAGE OF UK IMPORT PRICES BETWEEN 1951 AND 1963

	Freight % at highest price	*Freight % at lowest price*
Hemp (Manila)	8	15
Copra (Philippines)	7	22
Rubber (Malaysia)	2	5
Tin (Malaysia)	2	4
Jute (Pakistan)	5	11
Sisal (E. Africa)	4	13
Cocoa beans (Nigeria)	1·3	4·7

Source: UNCTAD (1968), based on information from Royal Netherlands Shipowners Association (1964).

the freight rates as a percentage of the market values of various commodities during periods of high and low prices between 1951 and 1963.

In fairness to the liner conferences it should be pointed out that freight rates frequently are reduced when commodity prices remain low over a long period. The mechanism for so doing is cumbersome: a shipper must make the request to the conference, not to the individual shipping lines with whom he does business, and the conference administration then deals with the request. Because many depressions are short-lived, and as adjustments to rates are usually made annually, a favourable response may not always be made when it is most needed.

On a more positive note, we can be reasonably certain that the export potential of the developing countries is quite extensive provided raw materials and semi-processed goods can remain competitive with synthetics, and trade liberalisation proceeds in relation to new manufactured products. In these respects the trends in shipping at least are favourable: over the years there has been a trend towards cheaper transport per ton of raw materials with the use of bulk carriers and specialised vessels, and the conference policy of charging lower rates on low-valued goods may, likewise, have assisted exporters in some developing countries, in spite of the misgivings about who bears these costs. Conferences have also from time to time carried new manufactured articles from developing countries at promotional rates to assist the build up of trade, as in the case of the exports of bicycles and sewing machines from India and Pakistan.[7] It must also be borne in mind that even when lower transport costs, as a result of technical advances in shipping, are not reflected in higher f.o.b. prices they may be manifest in lower c.i.f. prices, which thereby stimulate additional demand for the exports of developing countries.

Freight rates and imports
Since the industrial revolution in the countries of western Europe the products of industry have found outlets in Africa, Asia and other economically retarded areas. The impact of manufactured goods on many of these countries included the destruction of small-scale village crafts, and the stimulation of production of raw materials for export in order to purchase factory-produced goods. Since the early period many of the countries of the developing world have become economically dependent on the imports of manufactured commodities.

The rising prices of manufactured products exported from the developed nations have undoubtedly contributed to the widening of the gap in terms of trade between developed and less developed countries. Prices of goods produced in Britain rose by almost 4% per annum between 1959 and 1969, and at a higher rate thereafter; and export commodities of manufacturing countries being generally of high value carry high freight rates, which also have tended to increase.

The extent to which the producers of manufactured goods can transfer freight rates to the buyers of their goods depends on the degree of competition between companies and nations in any market. Generally the manufacturer sells his goods c.i.f., that is he quotes a delivery price to the buyer. He can therefore nominate the ship, but must gauge how much of the freight rates he can safely pass on to the importer.

For these reasons it is difficult to generalise about the effects of freight charges on imports on a developing country's economy. It could be argued that high freight rates simply add to the income gap and thereby render trade more difficult; it can also be argued that high freight rates have the same effect as tariff barriers in that they improve the competitive position of the developing country's domestic industries. In any event it is likely that when competition for markets is rife between the developed countries the exporters of manufactured commodities will absorb a high proportion of freight rates and will become active in cooperating with shipping companies in finding ways in which to reduce these rates. Unfortunately, containerisation, the main means being adopted to combat rising costs in the carriage of manufactured goods, cannot always be introduced to the trade of developing countries due to the inadequacy of supporting infrastructure at ports and in the hinterlands.

The drive for national shipping

The summary of the export and import situation in developing countries may have highlighted something which is not generally recognised, namely that the debate on freight rates is less of a conflict between shipping companies and the developing nations than it is between buyers and sellers of goods on a world scale. Increases in freight rates may be justified, by rising costs in ports, or some sector of transport. The real conflict lies in how the increases should be apportioned and this has little to do with shipping companies as such.

The debate on freight rates does, however, obviously affect the shipping industry. One of the reasons behind the drive towards more national fleets lies, in fact, in the belief that since a developing country has to bear the cost of carriage it will be less of an economic encumbrance if the freight payments, which are frequently deducted from the price paid to exporters, and sometimes added to the price paid by importers, are credited to national rather than foreign flag tonnage. In effect nationally owned ships would appear to ensure that a developing country receives the market price for its product in overseas currency and not the market price less freight charges.

The basic aims of national shipping projects are thus to increase export earnings and decrease spending on imports. But in any country every proposed investment project has to compete with many alternative uses for capital, while the relative values of investments are frequently measured in terms of social as well as economic cost benefits. Shipping may be more difficult to evaluate in these terms in a developing country than say a highway or a hospital, due to lack of data, such as shadow prices, and lack of experience in investment planning. But the addition of overseas currency as a result of investment in shipping may justify the economic and social benefits foregone in order to develop a national fleet. Furthermore, since as much as 80% of the cost of a ship may be borrowed from the main shipbuilding nations, and as this loan may not be available for other purposes, it can be argued that opportunity costs need be calculated only on the 20% down-payment.[9] The down-payment is the net immediate drain on the economy. This amount would be covered early by the annual net contribution resulting from the investment, after paying the loan reduction and the interest; that is, the investment will pay for itself and contribute to the balance of payments. These considerations make shipping enterprises attractive.

Other factors which may reinforce the decision to acquire a fleet include the possibility of expanding the quantity and type of exports. There is some evidence that trade can be generated by forging new links using national shipping. The Japanese did so with subsidised vessels during the early part of this century in a deliberate policy of trade penetration.[10] More recently Israel, Poland and Yugoslavia have opened wider trading relationships using their own shipping.

The national fleet as countervailing power to the conference also provides an incentive to purchasing vessels. However, with proportional representation in conference policy-making bodies it is unlikely that a developing country with a small fleet would carry much weight. Of even less real significance is the view that a national fleet provides outlets for labour. The capital/labour ratio is high in modern shipping and unless there are strong forward and backward linkages, by the creation of shipbuilding and ship-repairing for example, there are minimal employment generating effects.

Other incentives for pursuing the aim of self-sufficiency in shipping have been referred to in Chapter 5. They include earnings from cross trades, defence and strategic considerations, prestige, and the economic corollary to political independence. A more cynical view is, that since shipping can be developed much quicker than some other industries, and can therefore provide tangible evidence of development plan successes over a short period, it will receive high priority.[11] By the same token, as a ship is infinitely more easily obtainable and disposable on the world market than say a factory or a dam, it is less of an investment risk and thus justifies priority.

The combination of economic and non-economic incentives in the drive to introduce national fleets is certain to bring more ships of the developing countries into world trade; although the opportunities for credit from the developed countries have been reduced since July 1969, when the OECD group (which includes the major shipbuilding nations of Britain, Japan, West Germany, Sweden, Italy and the Netherlands) signed an understanding on export credits which was aimed at reducing soft shipbuilding loans supported by governments. The member countries also agreed on a maximum duration of payment of eight years from time of delivery, a 20% down-payment on delivery, and interest rates of not less than 6%.[12] Shortage of capital, harder credit terms, and the possibility of serious overtonnaging in world trade may slow down, for a time, the rate of progress in the building of new national fleets. It is likely, therefore, that the overall percentage share of the developing countries in the world fleet will remain relatively low, although the effects on specific trades could be significant. A brief review of activities in this respect may be useful.

Development of national fleets

Successful national shipping enterprises have been established in west Africa. In 1957 the government of Ghana obtained the financial and managerial assistance of the Israeli Zim shipping company in setting up the Black Star Line, and they entered the trade between Ghana and north-west Europe, and Ghana and North America. By 1960 the company had bought out the Zim interests and was a nationally independent, but entirely government owned, shipping line, the declared aim of which was to obtain 40% of ocean traffic between Ghana and the rest of the world by 1970.

The Black Star Line has been supported in trading by the government of Ghana, and by produce marketing organisations which have directed cargoes to national ships. Political problems in west Africa led to a 40% fall in cargo tonnages between 1965 and 1967 and to a general overtonnaging of the trade, but by 1969 the line was operating sixteen cargo liners within the West African Lines Conference (WALCON).[13]

The Nigerian National Shipping Line is also a member of WALCON. Indeed, the company was set up in 1959 with the help of the conference. The Nigerian government held 51% of the shares at the outset, the Elder Dempster Company 33% and the Palm Line 16%. The latter companies also provided managerial assistance. As with Ghana the foreign shipping interests were bought out in later years and the line became truly national in 1961.

The west African national fleets have made inroads into British liner services to west Africa. In 1952 UK liner operators controlled the whole of the west African trade; by 1968 they had only two-thirds of it. In the Rochdale report it is pointed out however that only part of this decline in the participation of British shipping was due to the growth of national fleets, the greater part was a result of the introduction of third flag operators.[14]

An impressive shipping development with less direct government participation applies to India. In 1969 the Indian fleet amounted to 3·3 million dwt, a growth of almost 70% since 1964. This rapid build-up of shipping is part of a development plan aimed at carrying 50% of India's overseas trade in Indian ships. The government has supported shipping with loans and rebates but does not provide operating subsidies.

On specific trades Indian ships had by 1965 obtained a substantial proportion of the available cargoes. In that year they carried 42% of all cargoes under the India-UK conference, and 39% in the India-Continent conference. Rochdale again records that, whereas before the Second World War UK ships carried four-fifths of the liner traffic on the Calcutta-US route, they had entirely withdrawn from this service as early as 1959.[15] In this case the abandonment of a traditional trade link was due mainly to the reservation of American-aid cargoes for United States and Indian vessels; the result was that foreign flag shipping carrying cargoes from India to the United States could not always secure back-loads and was therefore operating at an economic disadvantage. In the new trades which India has established to the USSR and Poland, Indian shipping has been even more favoured through bilateral trade agreements which ensure that half of the cargoes are carried by national flag vessels.

Pakistan, likewise, possesses a fleet which is growing rapidly. In 1969 it comprised 770,000 dwt, an increase of over 90% since 1964. The principal owner is the government-supported National Shipping Corporation (NSC) with about 300,000 dwt in 1969.[16] This body was founded in 1964 with 25% government shares and the balance raised from private investors.

The need for a substantial Pakistan merchant fleet was apparent on the formation of a fragmented state which comprised east and west provinces separated by a potentially hostile India. Whereas India tended to rely on railways for east-west trade (the distance from Bombay to Calcutta by rail being half the sea distance) Pakistan had immediately to establish sea links. Another stimulus to Pakistan shipping was the number of Moslem pilgrims who wished to travel to Mecca during the Haj season. The Pakistan Pan-Islamic Steamship Company now operate four passenger vessels between Karachi and Jeddah, and in 1967-8 carried 36,000 people.[17]

The aim of Pakistan, like that of India, is to acquire for its own ships a high proportion of the trade to and from the country. In 1971 the NSC had thirty-seven vessels on order, some being built in Yugoslavia on generous loan terms and others in the shipyard at Karachi. As far as can be determined from reports the ships of India and Pakistan are operating profitably within the conference. It also appears that participation in conference has brought with it opportunities to establish more favourable freight-rate terms for certain exports from these countries.

The state of Israel, like Pakistan, developed a national fleet partly in response to changing geopolitical conditions. When the state was established its land frontiers were closed and the only means of foreign trade was by sea. Cargo ships were chartered, and later acquired, and so also were old passenger vessels which, along with cargo vessels, helped bring almost one million immigrants into the country. The new state had thus immediate maritime orientations. In 1954 Israel received funds from Germany as restitution and with these new vessels were purchased. By 1969 the national fleet reached almost 2 million dwt and it carried about 50% of the country's seaborne trade. It was strongly supported by government funds throughout; however, the trend is now towards a fleet operating on a strictly commercial basis.[18]

Other countries in the Middle East are developing national fleets. Kuwait has an expanding fleet of oil tankers, the personnel for which is being trained by major oil companies. In Egypt private shipping companies were nationalised in 1961 and the United Arab Maritime Company was established. Between 1961 and 1969 the fleet increased only modestly from 201,000 to 239,000 g.r.t., carrying each year about 5% of the country's annual exports, about the same percentage of imports, and half of the coastal traffic. The most important of the latter was crude oil from terminals in the Red Sea to the refinery at Suez. Payments to foreign shipowners for this internal transport activity totalled £2·5 million between 1965 and 1968. In 1970 the Egyptian government laid emphasis on merchant fleet expansion in their Investment Plan 1970–5. It was calculated that even allowing for a low return on capital, or running at an accounting loss, the cost of earning, or saving, overseas currency was lower in shipping than it was in many other sectors of the Egyptian economy. Balance of payments considerations appears, therefore, to be uppermost in Egypt's decision to expand shipping.[19] A decision which was reinforced in 1971 by negotiations with Libya and Syria relating to joint Arab shipping enterprises for the carriage of oil in international trade.

Another major area of national shipping development is South America. Sea transport is at present the only economic method of commodity movement between the eastern coast countries (Argentina, Brazil, Uruguay, Paraguay), the western coast countries (Chile, Bolivia, Peru) and the northern countries (Ecuador, Colombia, Venezuela). Within South America the Latin America Free Trade Association has encouraged the development of shipping in order to widen markets for member countries.[20]

Brazil is the principal shipowning state in Latin America, with almost 2 million dwt in 1969. The fleet is mainly government owned. Next to Brazil comes Argentina with 1·5 million dwt in 1969. The only other important shipping nation in the area is Mexico with 570,000 dwt. Between them the Latin American nations have established a shipping capacity capable of conducting some overseas trading and a high proportion of intra-LAFTA trade, although trade among LAFTA members is only 10% of their combined exports.[21] Their system of mutual reservation of cargoes between the members, and flag discrimination in respect of overseas cargoes, could ensure the success of national and regional shipping. Reservation of cargo in South America has certainly had an effect on British shipping. Rochdale notes that in the UK–Uruguay trade the share of British lines was cut by half over the period 1962 to 1967, and in the UK–Brazil trade British shipping lost nearly a quarter of its normal cargo tonnage between 1957 and 1967.[22] The long-term aim of the LAFTA countries is to reserve over 50% of all imports and exports for their own ships. The realisation of this aim will require a substantial growth in the tonnage of shipping in South American countries.

In the other principal region of the developing world, south-east Asia, there is a number of growing fleets. Countries each with over one million tons of shipping include Taiwan, the Republic of Korea, Hong Kong and the Philippines, while Indonesia is approaching this level of tonnage.

Philippine shipping has been the subject of a consultant's report to UNCTAD.[23] This shows that the fleet comprises not only cargo liners, which are the main components of national fleets, but also several new oil tankers of up to 60,000 dwt, and specialised timber carriers. The principal ships of the Philippines fleet are, however, still the cargo liners; in 1967 they obtained 16·3% of conference cargoes from and to the Philippines.

The report on Philippine shipping provides evidence that for this country ship operating is profitable. Between 1964 and 1967 the average capital-output ratio was 2·25:1 which is very much better than the 5:1 to 7:1 quoted for most countries. In terms of balance of payments the Philippines fleet has made a progressive net contribution rising from 6·2 million pesos in 1963 to 93·3 million in 1967. These figures do not allow for the foreign disbursements in the Philippines had the national fleet not existed, nor for the import content of the fleet; but neither of these would substantially alter the positive balance of payments contribution made by the activities of the national fleet.

Other small fleets in the south-east Asia region include those of Thailand and Burma. The latter is a government-owned national company and was assisted in its inception by the Israeli Zim Line. The Chinese People's Republic appears, until recently, to have possessed only a small foreign-going fleet and to have relied on chartering tonnage from overseas shipping companies. Between 1965 and 1970 China's merchant fleet increased by 100% to reach 1 million g.r.t.

Finally, on a much smaller scale, national fleets are developing in the Pacific region. The isolated Gilbert and Ellice Island Colony (GEIC) has, for example, found it necessary to purchase a foreign-going vessel in order to maintain trade links with Australia. This is due to the remote location of the islands in relation to international trade routes. The Kingdom of Tonga which is geographically more accessible than the GEIC to international shipping has also established a small national fleet which includes one principal foreign-going vessel. Like many of the ships in the fleets of other developing countries this vessel was purchased second-hand. It has proved inadequate in design and speed for the Tonga–Australia trade, and this, plus bad operating practices, has made it difficult for the ship to pay its way.

Comments on national fleets

It seems likely that the bigger national fleets are profitable, but, as in some developed countries, subsidies, tax rebates, and reservation of cargoes are common. Whether or not the capital invested in shipping could have brought greater benefits had it been used in some other sector of the economy is open to debate. Certainly, the World Bank has, in the past, tended to treat shipping investment with some scepticism, for it has not encouraged this by providing loans. Mr H. J. Van Helden of the World Bank has pointed out that, while no foreign companies will invest in the roads and railways of a developing country, overseas companies will invest in ships and aircraft to meet a developing country's foreign trade needs.[24] The implication is that developing nations should first invest in their country's infrastructure and utilise the already available shipping services for their overseas trade.

The above reference by the World Bank representative to internal communications should also include ports and coastal shipping. Some of the high freight rates charged by liners may be attributable to port delays, a factor which will also raise costs in national shipping enterprises irrespective of cargo reservation and other measures. If, therefore, there is a diversion of scarce capital away from ports and internal transport improvements in order to duplicate international services this may be a doubtful use of resources. On the other hand there have undoubtedly been very substantial contributions made by new foreign-going national fleets to the economies of some developing countries, although measurement of overall economic gains is difficult to make.

The losses to a sector of the economy, the primary producers, from poor internal transport and port systems may be more easily assessed than the possible gains from national fleet developments. For example:

the Dundee Importers' Association quote ocean freight rates for jute carried over 12,000 miles as £19 per ton, compared with £4·13 per ton for a 200-mile haul to the ports in what was east Pakistan. In the Fiji Islands in 1964 the local shipping charge on copra was $12 per ton over a distance of 150 miles, which was about the same rate per ton as the transport from Fiji to Japan; and in 1968 in Thailand the rate per lb for rubber was 1·71 US cents for local carriage and handling, and 1·373 US cents for sea transport to Europe by conference line.[25]

While national fleets may improve the balance of payments, they may not always represent the best use of resources in terms of raising the overall living standards of primary producers, if this is sought. On the other hand, the evaluation must in all cases be carried out in terms of social cost benefit analysis. It is reasonable to assume, however, that improvements in internal transportation will bring benefits to primary producers by reducing the differential between f.o.b. prices at the ports for products and cash returns to producers. Even if the efficiency of foreign-going shipping, national or otherwise, is greatly improved some of the gains can be dissipated by poor internal distribution and collection. In this respect the coastal and short sea trades of many of the developing areas have a very important role to play in the national economy. A review of shipping would be incomplete without reference to the domestic shipping services of some of the developing countries.

Types of coastal and short sea shipping in developing areas
There is great diversity in the craft employed on coastal and short sea services in developing areas. The physical and economic conditions of operation are even more varied, but there are some common characteristics. These will be outlined before considering specific examples.

Generally, vessels divide into two main groups. First, the cargo/passenger ships of from 200 to about 1,000 dwt. Most of these have been purchased second-hand from countries in north-west Europe where they were usually employed in the carriage of coal, timber and fertilisers. They are not always suited to the carriage of tropical products, such as copra, hemp, or rubber which have high stowage factors; for on loading these lighter commodities the holds are full long before the deadweight capacities of the ships have been reached. The earning potentials of vessels transferred from northern Europe to the tropical world are not therefore always fully utilised. The variety of ships bought from other areas means also a lack of standardisation in spare parts, which makes stockholding and repairs expensive.

The second group of vessels include small wooden craft such as cutters, schooners, dhows, and canoes. These are frequently locally built; but in design, and often in age, they relate to an earlier period of trading when many of the developing countries were part of the colonial empire of one or other of the major maritime powers. Usually the resident merchant companies of the metropolitan countries operated their own ships on the principal domestic trade routes, leaving the remoter and more difficult areas for indigenous entrepreneurs. Now most of the developing countries must themselves provide all internal shipping services.

In investing in shipping local companies in the developing countries have usually given priority to the purchase of vessels for employment in regions which, traditionally, have generated most of the trade. Many of the vessels bought from foreign sources were not in good condition, but with a shortage of capital and overseas exchange difficulties this was unavoidable. As a result of the initial capital costs and the continuous repair costs of craft bought overseas it has been even more difficult to obtain finance to replace the older types of vessels engaged in servicing the less productive and remoter areas. Many of these vessels, which are twenty to fifty years old, have deteriorated through lack of maintenance and due to the general corrosive effects of operating in tropical waters.

Some of the older and smaller craft are well suited to the physical environments of outer areas, but they are not well suited to the socio-economic conditions that have evolved in recent times. Rates of crew wages have risen, and so have standards of living and safety provisions in the maritime industry generally. These trends render small labour-intensive craft less economic. In the outer areas too, many people expect better standards of cargo services than previously, and they demand more safety when travelling on vessels.

The infrastructure supporting local shipping is, like the vessels, very variable in quality between the regions of a developing country. Main channels may be reasonably well buoyed for day and night navigation, but away from the main routes even daylight navigation can involve picking out passages between reefs and coral heads—often with no assistance apart from visual observation from the masthead, and night navigation is often impossible.

Port facilities are likewise primitive in many areas. At each end of the principal domestic trade routes there may be wharves to which roads are connected, but on other stretches of coastline settlements are without landing facilities and they are seldom linked laterally by proper roads. The result is that vessels working along a coastline have to anchor, or stand off and on, at almost every settlement, and send their boats in to deliver and collect cargo. The boats may have to negotiate difficult reef passages, wait for tides, to shoot reefs on the crests of waves, before they can reach the villages. Under these conditions voyages are protracted and costly for small quantities of cargo.

The combination of poor-quality craft, inadequate infrastructure, and the dangers inherent in trading under difficult physical conditions, render local shipping unattractive to private investors. There is often a vicious circle. To enable vessels to break even financially, owners avoid costly repairs, employ poorly trained personnel, and tend to overload with deck cargo and passengers whenever possible; also the captains often try to push the vessels along at night in dangerous waters to save time. Accidents and losses are thus prevalent. To minimise the financial risks involved owners frequently employ several small very undercapitalised craft, with the minimum of safety provisions, rather than quality vessels properly equipped and manned. This has the effects of a high accident rate and a low rate of private capital flow into the maritime industry.

Generally, governments do not support their local shipping financially for they fail to see these as earners of overseas currency. In fact domestic shipping appears as a drain on overseas reserves through the purchase of vessels and spare parts abroad. This attitude arises from a mistaken view of local shipping as a single entity rather than as part of a chain of transport which serves the export industry, and which may act as a stimulant to the social and economic development of the country.

The role of domestic shipping

The numerous inadequacies of domestic shipping in some developing countries would be of little consequence were the people in the outer areas still living in a state of primitive affluence, growing their own food, obtaining clothing and housebuilding materials from the immediate area, and generally meeting most other cultural and economic needs by barter and other forms of indigenous trade with similar close-knit communities. This is no longer the case. Even very remote societies have been undergoing processes of relatively rapid cultural and economic change, and they now rely on, and desire, a considerable range of the commodities considered as essential for modern society. What is more, the governments have encouraged production of cash crops for the export market in support of the national economy, and have done so by stimulating new needs.

The coastal, riparian, and island communities in developing countries thus depend on local vessels to keep stores and cooperatives supplied with basic necessities, to provide carrying capacity for products and livestock to reach national markets and entrepôts, and for the provision of travelling facilities for passengers, and also, in some areas, for medical and administrative services. If domestic shipping fails to carry out all of these functions reliably, and at reasonable cost, then the outer area communities are placed at an even greater economic disadvantage than ever before when compared with those located nearer the main port towns, and these towns are the growth points in many of the developing countries. The role of domestic shipping is thus basically to help overcome the dichotomy in cash incomes and opportunities between urban growth points and rural areas.

The port towns and cities from which domestic ships operate to outer areas are the places from which the processes of social and economic change radiate. Many of these towns appeared during the colonial period as foreign enclaves at which overseas administration, merchant companies, banks, and medical and educational facilities were concentrated. Most of the cargoes arriving from and destined for overseas countries were channelled through the ports, and it was there that harbour improvements and wharf construction took place. Road building funds and most of the other social overhead capital expenditure of the colonial period was also directed to areas in and around these principal ports of entry, and it was in this growth environment that ship repair facilities, and some manufacturing and industrial processes, developed.

In recent years the port urban centres have everywhere burgeoned in population. People have arrived from the countryside and outer areas seeking wage-employment, educational facilities for their children, or simply an escape from the backwardness and constraints of traditional village life. The dense concentrations of population and the relatively dynamic economic environments of the growing towns have made them the most obvious places for further investment in manufacturing and industry. The towns are the places at which a high proportion of a country's scarce capital must inevitably be allocated for more roads, schools, hospitals, water supplies, sewerage, and power supplies.

The continuous concentration of new investment at the principal towns makes them increasingly attractive; but this growth can produce, in Myrdal's terminology, either backwash or spread effects amongst the hinterland communities.[26] Backwash effects may be said to predominate where contacts with the port centres are spasmodic, where marketable resources are idle, and where there is a net migration, especially of young people, to the towns. The spread effects may be said to outweigh the backwash effects in areas where there are regular transport facilities for contact with the towns, where the area's potential for market production is being fully realised, and where per capita cash incomes show cumulative increases. There is thus an implied relationship between the costs and efficiency of the transport system linking the ports to the outer island and coastal hinterlands, and the degrees of spread or backwash effects experienced in these hinterlands.

Marked regional differences in incomes in a developing economy can, of course, be ascribed to numerous factors other than transportation problems. However, improved transport has always rated highly as a prerequisite for economic growth. Nowhere is this more true than in countries where there are under-utilised land and resources in relatively remote areas, and over-urbanised port towns in inner zones—which may even be importing foodstuffs from overseas. The socio-economic significance of port town–hinterland links by small vessels, and the problems involved in improving these, may be appreciated from a few regional examples drawn from the Pacific and south-east Asia.

Pacific inter-insular trades

The Pacific islands have been aptly described as sea-locked countries. They comprise atolls, reef islands, high volcanic islands and high continental islands. Several are extremely isolated from international shipping routes, almost all are fringed by coral reefs, and many have off-shore barrier reefs.

Pacific seamen distinguish the following types of coral islands on the basis of their accessibility to trading vessels. Lagoon islands where the reef and land area encloses a lagoon which can be entered by an inter-insular trading vessel; outside anchorages, where the reef and land area encloses a lagoon which can be entered by boats only; and reef islands, where there is no lagoon and the fringing reefs are steep-to, so that an anchorage is unsafe and the ship must lie off the island.

In Micronesia and parts of Polynesia the principal ports have developed at lagoon islands. In the Gilbert and Ellice Island Colony, for example, Tarawa atoll with its lagoon of 132 square miles is the location of almost all administrative and commercial facilities. From this port a scatter of small islands, extending over almost 1,000 miles of the Pacific Ocean from north to south, is served. At Tarawa all cargoes from and to overseas countries are unloaded and loaded, and distribution and collection is made throughout the whole of the group by a few local vessels.

In this example, where the population numbers less than 50,000, and where cargoes from and to the islands are in minute quantities, the government has to provide financial support for sections of the local shipping. Also, to enable ships to operate to very distant islands, which because of population pressures and difficult environmental conditions produce only a few tons of copra annually, a standard freight rate policy has been adopted. This involves the islands near to the main port subsidising the more distant places. The result is that the nearer islands are charged three times the freight rates which they normally would pay, and the very distant islands pay about three times less than they would be charged on a cost of carriage basis.

The uniform price system thus attempts to counteract the trends towards inequalities of income which result from locational disadvantages in relation to a growth point. On strictly economic ground this policy may be difficult to justify, for as Bauer has pointed out in respect of west Africa, a uniform price system results in the use of greater amounts of scarce resources to yield a given total output of the crop, and brings about a penalisation of some producers for the benefit of others.[27] It may, however, be justified on social ground since people in the distant areas would receive very little for their produce were the marginal cost for transporting these charged. In the Gilbert and Ellice we can thus clearly see how the whole economy is burdened by geographical fragmentation of the territory and the vital role which shipping has in knitting the parts into an economic whole.

A somewhat different system of inter-insular trading operates in Fiji. Here there are many rich high volcanic islands with greater potentials for agricultural production than exist on the low coral atolls. The port town Suva, with its excellent sheltered harbour, dominates the archipelago in terms of population size and multiplicity of functions. It is from Suva that imported goods are distributed to most of the outer islands and it is to Suva that island commodities are transported for processing and for export. The local vessels range from small cutters to ships of over 200 tons. They are almost all privately owned; but the owners can obtain financial returns only by carrying excessive numbers of passengers, in addition to cargo, curtailing maintenance and repairs, and charging very high freight rates.

The practice of working around many settlements in the outer islands of Fiji means that a small cutter can spend four or five days unloading and loading a total of 40 to 50 tons of cargo. Not only does this reduce the vessel's earning potential but it means that fresh produce, fish and livestock, cannot be safely carried, so that resources of the rich volcanic islands are often wasted. Returns of producers in the remote islands for their copra are also lowered due to the high freight rates. In 1964, for example, the freight rate for the haulage of copra 200 miles from Kabara Island to Suva was $13 per ton, by contrast the freight rate per ton of copra from Suva to Japan (about 4,000 miles) was only $11·5.

Fiji epitomises the problem of the spatial fragmentation of an economy, and also the social role of domestic shipping. This may be illustrated as follows: many of the villages in the outer islands of Fiji and those in remoter coastal locations on the main island, have a one-stepped relationship with the dominant port town of Suva. That is, each has a small village store which, because of its limited market, can stock only basic goods such as flour, sugar and cigarettes. To obtain goods of a slightly higher order it is necessary for trips to be made to shops in Suva, for there are no bigger shopping centres lying between the village and the city. If villagers in remote parts of Fiji possessed intervening commercial opportunities for obtaining a greater range of goods at an island or regional centre then the cost of fragmentation might be reduced. At such places the shops would have the total population of the region as a market and much needless travel and expense in reaching the port town would be avoided. If, in addition, such a centre had a small wharf, and local people and their vessels, or road vehicles, were used to concentrate cargoes from the surrounding settlements at that point, then bigger ships could be employed on regular services from the port town to one point only, thus solving some of the problems of shipping viability. Villagers would, as a result, have a two-stepped relationship with the capital, day to day goods would be bought at village level and higher order goods at the regional centre, at which their copra, fresh produce, and livestock would also be sold.

Rationalisation of the above nature would require changes of a social as well as an economic kind. In developing countries social traditions are deeply ingrained, and in the island world shipping plays an important part in the social networks. People have, for example, moved in great numbers from remote areas to the port towns, and there they have frequently formed residential groupings according to places of origin. They maintain links with their home area and kindred through local ships, and these kinship ties generate a considerable movement of passengers and interchange of gifts. Thus the social role of shipping within the existing spatial pattern of village/town shopping facilities is obviously important. The assumption that regional centres might reduce needless movement of people from and to the port town is, in these circumstances, not necessarily true. Regional centralisation in a fragmented economy would, however, allow more viable transportation services.[28]

The islands of east New Guinea may illustrate yet another aspect of the importance of local sea transport, the need to raise this to higher standards and, at the same time, reduce costs of operation. Up until the 1960s foreign-going ships operating from Australia called at many of the ports in the outer islands of New Guinea. With the trend to srteamlining foreign-going services by reducing ports of call it was left to local craft to serve the outer ports and to act as feeders to principal ports. They were inadequate for this task and could carry it out, in a haphazard way, only by charging very high rates; so that the loss of the foreign-going services resulted immediately in a reduction in the net received price of products by $5 to $20 per ton at various places.[29]

Considerable emphasis has been placed in the development plans of Papua–New Guinea to achieving an adequate internal transport service closely coordinated with the programme for primary industries. This is of vital importance in a country where people are being induced to grow new types of crops for sale on the world market. The risks involved from the price instability of primary products are great and only a little can be done to shield producers from these, but at least the internal shipping can be provided as a service to, rather than as a drain on, the producer; if the correct types of vessels are adopted and optimum patterns of trading established.

Inter-insular shipping in south-east Asia

The domestic fleets in Indonesia, Malaysia, and the Philippines are on a very much bigger scale than in the south-west Pacific.

Indonesia comprises 3,000 islands extending for over 3,500 miles in an east-west direction. The total population of the country is around 100 million; about 60 million of which is concentrated on the very fertile island of Java and the rest widely scattered. The people in the outer areas are culturally and linguistically diverse and have no obvious affinities with the main island of Java. Good inter-island communications have always been of political as well as economic significance in this region.

Under Dutch colonial rule the archipelago was served by the highly efficient inter-insular ships of the Royal Packet Navigation Company (KPM). They had a virtual monopoly of the main routes but the trade of very isolated islands and small settlements was conducted by local sailing perahus.

Since independence Indonesian domestic shipping has passed through several phases. The Dutch fleet was withdrawn between 1955 and 1957 when foreign company operating licences were revoked. Its place was taken by a variety of second-hand ships purchased from the countries of eastern Europe, the USSR, Japan and Hong Kong, and by an increased fleet of perahus locally built. There is no doubt that for some places this policy was disastrous since neither foodstuffs

Shipping and the developing countries

nor raw materials could find adequate tonnage, due to the dual defects of attempting to operate inferior shipping with inexperienced personnel. In recent years the position has improved. Although official statistics are not available it seems likely that the main fleet of inter-insular ships in the mid-1960s numbered about 200 vessels, aggregating 400,000 dwt. There was probably about the same aggregate tonnage of perahus.[30]

The basic transport task in Indonesia is that of linking economically and politically the centres of dense population with the less populated food and raw material producing islands. Many of the latter export their timber, rubber and other raw materials direct by overseas tramp ship from regional ports, and oil is exported from Sumatra and Borneo by tanker. By contrast the ports of Djakarta and Surabaja in Java import almost all consumer goods, although they manufacture increasing quantities of these. The domestic fleet delivers consumer goods to the outer areas and returns with rice, sugar, maize and other foodstuffs. If the ships fail to operate successfully and economically it can be serious for Java's massive population, and can create disaffection in the outer areas where people are well aware that much of their raw material supports the total economy, and their foodstuffs production helps sustain Java. This disaffection is added to by the fact that the outer islands are prohibited from importing, by cheaper means, some consumer goods direct from overseas and must rely on costly transhipment from Java.

As most of the prospects for further production of food and raw materials lie in the outer areas an improved inter-insular fleet has an important role to play. This was recognised in 1963 when the Zamen Bahari—'Age of Sea Glory'—was declared, under which the expansion of national shipping would bring Indonesian vessels back on the high seas, and also integrate the archipelago into one nation. However, with a shortage of capital there has been some sacrifice of the vital inter-insular shipping in favour of foreign-going vessels which have more prestige value.

It is interesting to compare Indonesia with Malaysia. The latter has not developed a foreign-going fleet of any size; probably because of the proximity of the Strait of Malacca, and the port of Singapore, on which focuses most of world shipping trading to the Far East and south-east Asia, thus offering considerable export capacity. On the other hand, marketing organisations in Malaysia have been very active in negotiating favourable freight-rate terms with the conferences.

The inter-insular shipping in the area focuses on Singapore, from which port domestic trade routes radiate to the east and west coasts of Malaya and to Sarawak and Brunei. As in Indonesia overseas ships load timber, minerals, and oil direct from ports and anchorages along the coasts, but local vessels are active in collecting smaller shipments of rubber, copra, hemp, palm oil and cocoa for transhipment at Singapore. There is less emphasis on the carriage of foodstuff than in Indonesia, for Singapore is normally well supplied with this from the immediate hinterland. Cargoes carried outwards in the local trade from Singapore, and to some extent from Port Swettenham, comprise a vast variety of manufactured goods and equipment.

An interesting account of decision-making related to the type of technology to adopt to achieve improvements in the above trade has been given by Mr Robin Postlethwaite of the Straits Steamship Co. Ltd, Singapore.[31] The first problem they were faced with was whether to introduce unitised vessels with automated cargo-handling equipment in place of conventional ships. The direct financial saving in port costs by so doing in a region of low-cost labour was seen to be slight; but the saving in time for the ship appeared to make unit-load operations viable. The next process was to decide on the type of unit vessel and methods to adopt.

It might have seemed that in an area with 18 or so ports along a coastline there was a case for the Lash ship. But the capital costs of this system and the need to provide tugs to marshal barges in the ports ruled it out. The roll-on/roll-off technique was also unsuitable since there were few road systems which could be used to achieve through-transport with motor vehicles. The so-called Combo ship seemed possible since it could combine features allowing roll-on/roll-off at some ports, pallets and pre-slung cargo at others, and the carriage of containers as well as conventional break-bulk loads. The disadvantages were costs of such an elaborate ship, its low resale value, and the slowness of turnround with some categories of cargo, which would be uneconomic with a capital-intensive vessel.

The remaining alternatives to consider were pallets and containers. High rainfall and lack of transit sheds in outer areas placed pallets at a disadvantage, whereas the container was rainproof and did not require storage facilities. Also, with so many ports to cover, the container could be unloaded at any time and left on the wharf for the contents to be discharged while the ship proceeded along the coast. The decision was, therefore, to adopt a container system.

The adoption of a highly capital-intensive system in an area with relatively primitive port facilities may appear surprising. But ships fitted with cranes, and carrying simple trailers and prime movers for use at each port, proved possible. Two such ships of $17\frac{1}{2}$ knots with container capacities of about 440 were ultimately selected to replace ten older conventional vessels.

The container decision had two additional advantages. First, it was in line with plans at the main ports for container facilities to handle overseas cargoes, second because of the imbalance of cargoes in favour of outward loads from Singapore it offered safe carriage facilities for fragile manufactured wooden dowels and mouldings, etc, to be transported from the timber-producing areas to Singapore, thus it could help diversify and raise the value of outer area exports. It would also, thereby, make possible the spread of industrial processing from the port town to the outer areas.

In other parts of south-east Asia development of the inter-insular fleets is equally important and investment decisions are as crucial. The Philippines, for example, show great unevenness in the distribution of population over 880 islands, and a characteristic concentration of commercial activities at the port-town entrepôt of Manila which handles about 90% of overseas imports but only 30% of exports of copra, sugar and timber. In the Philippines much of the inter-insular fleet still comprises wartime US landing craft, competition is intense and it is difficult to make the vessels pay. As elsewhere capital is required to be raised for purpose-built vessels suited to local needs.[32]

In Burma a variety of small craft ply the 4,000 miles of navigable waterways and provide the lifeline of riparian communities. This is true also of Thailand, particularly along the River Menam and the delta areas, but in most cases these services are starved of capital and are highly accident prone. Other regions in which coastal and short sea vessels play an important role include the Persian Gulf–Red Sea–east Africa area, and the Caribbean.

In many areas of the developing world local sea transport will remain a vital element in economic and political geography. Low-grade and costly shipping, or unreliable schedules, can discourage production for the market. Conversely, services of a high standard can help generate production by reducing the time and cost barriers in the local economies.

What type of shipping is best adopted depends on physical, economic and cultural conditions. In some areas small craft sailing regularly to the port towns act as economic stimuli, for while they may appear overmanned and under-capitalised the people using them may be producing marketable goods simply to justify visits to the main centre; if deprived of this opportunity by bigger commercial vessels, with less frequent schedules, production could decline. The local boatbuilders may in turn be employing plentiful local timber and labour resources and thus conserving scarce capital which would otherwise have to be employed for purchasing vessels. On the other hand, a more sophisticated form of transport, using relatively big vessels with refrigerated space, could release under-used local resources. There are few general rules in these respects.

In many cases it is the trends to containerisation and specialisation in the wider international maritime linkages which are now influencing regional shipping. The latter have almost always acted as local extensions of the former, so that with the technical and routeing changes coming about in international sea transport there must come complementary changes in local transport systems. It is interesting to contemplate that this process is part of the age-long transmission of cultural, as well as economic, traits along maritime routes. Moreover, it is remarkable that it is focal points in south-east Asia, the isthmuses of Suez and Panama, the Cape route, and even the Strait of Dover, which continue to exert strong influences on patterns of maritime transport and thereby on commercial and political affairs in the world.

1. Jacobson, H. L., 'Export Opportunities for Developing Countries', *Progress* **4** (1969), pp. 173–8.
2. Ho, A. K., *The Far East in World Trade: developments and growth since 1945* (New York 1968).
3. Richards, P. J., 'Shipping Problems of Underdeveloped Countries', *Bulletin Institute of Economics and Statistics* **29** (Oxford 1967), p. 265.
4. UNCTAD. Sarangan, T. K., *Liner Shipping in India's Foreign Trade* (UN, New York 1967).
5. UNCTAD TD/B/C.4/66. *Review of Maritime Transport 1969* (UN, New York 1969), p. 9.
6. Rochdale Report, *Committee of Inquiry into Shipping* (HMSO, Cmnd. 4337, London 1970), p. 119.
7. Naqvi, S. I. H., *The Operation and Practices of Shipping Conferences*, unpublished dissertation, (UWIST, Cardiff 1971), p. 20.
8. Kennedy, M. C., in Prest, A. R., *The UK Economy* (London 1970), pp. 34–5.
9. UNCTAD TD/B/C.4/85. *Development or Expansion of Merchant Marines in Developing Countries* (UN, New York 1971).
10. Furuta, R., and Hirai, Y., *A Short History of Japanese Merchant Shipping* (News Service, Tokyo 1967).
11. UNCTAD TD/B/C.4/32. Tresselt, D., *The West African Shipping Range* (UN, New York 1967), p. 48.
12. OECD, 'Understanding On Export Credits For Ships' (1970) in UN, TD/B/C.4/58.
13. UNCTAD TD/B/C.4/32, op. cit. Provides comprehensive background to west African lines.
14. Rochdale Report, op. cit., pp. 47–8.
15. ibid., p. 45.
16. *National Shipping Corporation of Pakistan Report 1968–69* (Karachi December 1969).
17. *The Pan-Islamic Steamship Co. Ltd. 18th Annual Report 1968* (Karachi May 1969).

18. Wydra, H. N., 'Israel Emerges as a Maritime Nation', *Journal of the Israel Shipping Research Institute* (March 1971), pp. 2-3.

19. Brown, R. T., *Transport and the Economic Integration of South America* (Brookings Institute, Washington D.C. 1966).

20. *UN Statistical Yearbook* (1969), Table 147 (Egyptian tonnages); and Ahmed Abulhassan Aly, *Analysis of Capital Investment in Merchant Ships in UAR (Egypt) during the period 1962-63,* unpublished master's thesis (Faculty of Commerce, University of Alexandria 1969), pp. 314-18, and unpublished work (UWIST, Cardiff 1971).

21. Mena, A. C., 'Latin America: the IBD', *Bolsa Review* 5 (August 1971), pp. 444-8.

22. Rochdale Report, op. cit., p. 45.

23. UNCTAD TD/B/C.4/84. Fabella, Armand, *The Development of the Merchant Marine of the Phillippines* (UN, New York 1971).

24. Van Helden, H. J., Statement to *Interregional Seminar on Containerisation* (UN Resource and Transport Division, London 1-12 May 1967), p. 4.

25. UNCTAD TD/B/C.4/85. *The Maritime Transportation of Jute* (UN, New York, 1971). UNCTAD TD/B/C.4/60/RWI. *The Maritime Transportation of Natural Rubber* (UN, New York 1970), p. 71.

26. Myrdal, G., *Economic Theory and Underdeveloped Regions* (London 1963).

27. Bauer, P. T., *West African Trade* (Cambridge 1954).

28. Couper, A. D., 'Rationalising Sea Transport Services in an Archipelago', *Tijdschrift Voor Econ en Social Geografie* (July-August 1967).

29. *The Economic Development of the Territory of Papua, New Guinea* (John Hopkins Press, Baltimore 1965).

30. Shamsher, Ali., 'Inter Island Shipping', *Bulletin of Indonesian Economic Studies*, 3 (Australia National University, Canberra 1966).

31. Postlethwaite, Robin., 'Case Study of Services in South-east Asia Trade', Interregional seminar on coastal feeder and ferry services (Norway September 1969).

32. Wernstedt, F. L., 'The Role and Importance of Philippine Inter-insular Shipping and Trade', *Data Paper 26* (Cornell U.P., Ithaca 1957).

INDEX

Aden, 26, 62
Africa, 26, 37, 38, 39, 40, 50, 70, 71, 76, 102, 104, 105, 110, 127, 133, 143, 145, 146, 172, 179, 182, 186, 201
Air transport, 73, 74, 83, 103, 104, 106, 107-8, 172
Aluminum, 132, 143-4
America, 42, 51, 75, 76, 92, 116, 141, 159, 161
Arabs, 26, 27, 33, 36, 37, 38-9, 40, 42, 62
Arctic, 60, 147
Argentina, 94, 188
Asia, 20, 37, 42, 159, 183
    south-east, 30, 38, 39, 40, 41, 42, 43, 45, 46, 49, 70, 76, 110, 112, 143, 179, 180, 189, 195, 199, 201, 202
Associated Container Transport Ltd. (ACT), 155
Atlantic Container Lines (ACL), 150, 153, 155-6
Atlantic Ocean, 26, 27, 28, 43, 50, 52, 54, 56, 58, 60, 63, 105
    north, 59, 67, 73, 106, 152, 154, 155
Australia, 40, 41, 46, 51, 54, 55, 62, 71, 94, 107, 133, 140, 143-5, 147, 150, 152, 154, 156, 157, 159, 164, 169, 190, 198

Balance of payments, 56, 79-81, 180, 185, 188, 189, 191
Ballast, 52, 72, 82, 91, 97, 136, 137, 145
    ballast voyages, 72, 93, 138
Baltic Exchange, 50, 55, 81
Baltic Sea, 28, 32, 33, 40, 45, 60, 61, 64, 95, 127, 159
Bantry Bay, 120
Barge-carrying vessels (BCV), 148, 151, 153, 154, 157

Barges, 148, 151, 160, 163, 164, 169, 176, 200
Black Star Line, 186
Brazil, 46, 51, 71, 145, 147, 188, 189
Brigs, 15, 47
Bristol, 45, 50
Britain, 28, 29, 30, 33, 41, 55, 64, 65, 76, 77, 78, 87, 95, 102, 104, 119, 133, 134, 136, 141, 146, 148, 150, 155, 176
    coastal shipping, 159, 160-2, 163-5
    development of shipping, 44, 46-54, 56, 69, 71, 90-1
    economy, 56-7, 80-1, 166-7
    passenger liners, 105, 106, 107, 108
    short sea trade, 166-9, 170, 171, 172-3, 174
    tankers, 115-16
Bronze Age, 25, 29, 30
Brunel, I. K., 54
Bulk carriers, 73, 74, 82, 86, 92, 104, 131-57, 168, 183
    cargoes, 140-4, 169
    dry cargo, 131-2, 138
    OBO, 137
    ore, 131, 133-6, 144, 145, 146
    UBS, 137
    unitised shipping, 147-57
Byzantium, 32, 33, 35, 36, 37

Canada, 51, 65, 71, 133, 141
Cape of Good Hope, 67
Cape route, 38, 42, 50, 60, 62, 63, 71, 114, 116, 117, 124, 125, 137, 141, 143, 202
Cape Town, 51, 52, 67
Caravel, 43
Cardiff, 91
Cargo-handling, 51, 103, 105, 145, 147, 153, 156-7, 162, 164

Index

equipment, 54, 74, 90, 98–9, 100, 132, 137, 143, 145, 151, 163–4, 174, 175, 200
Cargo liners, 37, 55, 73, 74, 84, 90, 98–105, 140, 186, 187–8, 189
   coastal, 161
   costs, 100–1, 152
   description, 98–9, 100
   liner market, 83, 92
   passengers, 99, 191
Caribbean Sea, 110, 201
Carracks, 36, 38
Celts, 30, 32, 33, 34, 35
Central America, 125
Channel tunnel, 177
Charter party, 15, 82
Charters, 28, 81, 82–3, 91, 165
   bulk carriers, 136, 138, 140, 144
   tankers, 119, 188
   tramps, 92–3, 95, 96, 97
   types, 82–3, 116, 118
Chemical carriers, 74, 128–9, 160, 170
Chemical industry, 83, 125, 129, 143, 165, 170
China, 28, 30, 38, 39, 40, 41, 42, 43, 51, 70
c.i.f. (cost, insurance, freight), 15, 96, 183
Clinker-built, 15
Clippers, 15, 51, 52, 54
Coal, 45, 47, 48, 52, 54, 56, 57, 63, 71, 73, 91, 96, 110, 116, 132, 134, 137, 138, 141, 143, 144, 161, 163, 168, 169, 191
   trade, 45, 46–9, 69, 71, 93–4, 141, 163, 165, 168–9
Coastal shipping, 28, 40, 125, 134–6, 146, 159–71, 190, 191
   British, 46, 47, 48, 57, 160, 161, 163–4, 165
   developing countries, 190, 191, 201
   vessels, 163–5
Coir, 15, 39
Coke, 133, 134, 161, 168
Colliers, 47, 48, 161, 163
Collision, 61, 65, 66, 87, 121, 122
   anti-collision tracks, 67
Combined carriers, 74, 82, 131, 136 137, 138, 140
COMECON, 64, 76
Conferences, 15, 83, 101–3, 155, 156, 182, 183, 185, 189, 191, 199
   conference system, 101–2, 104
   India–UK conference, 102, 187
   WALCON, 186
Consortia, 148, 153, 154, 155
Container ships, 63, 64, 74, 148, 149, 151, 152, 153, 154, 155, 156, 164, 170, 171, 173–4, 175
Containerisation, 152–3, 154, 155, 184, 201, 202
Containers, 147–8, 149, 150, 152, 153, 157, 164, 170, 171, 173–4, 175
Copra, 96, 179, 191, 196, 197, 200, 201
Cranes, 137, 143, 148, 149–50, 151, 153, 164, 171, 200
Cruise ships, 107–8
Curragh, 29, 33

Denmark, 29, 32, 66, 163, 177
Depths, 61, 62, 114, 120, 134, 144, 145, 146, 148
Developing countries, 72, 75, 78, 81, 100, 104, 153, 157, 179–202
   domestic shipping, 191–201
   economic status, 179–81, 197
   freight rates, 180–4
   national fleets, 189–90, 191
   national shipping, 184–6
Dhows, 192
Draughts, 54, 61, 62, 63, 66, 98, 114, 120, 145, 148, 151, 165
Dundee Importers' Association, 191

Economies of scale, 86–8, 92, 98, 129, 132, 145, 146, 152–3, 156, 179
EEC, 64, 144, 167, 169, 176–7
EFTA, 64
Egypt, 23–5, 26, 36, 37, 41, 42, 66, 188
Eilat, 66
England, 28, 34, 35, 36, 46, 91, 161, 175
English Channel, 51, 66, 166, 174, 177
Entrepôts, 23, 24, 26, 27, 33, 36, 39, 40, 43, 120, 193, 201
Environment, 59, 61–2, 121
Etesian winds, 16, 24

f.a.k. (freight all kinds), 15, 176
Far East, 49, 63, 66, 107, 112, 154, 155, 157, 179, 199
f.d. (free discharge), 15
Federal Maritime Commission, 155
Feeder fleets, 40, 164, 196–8
Ferries, 161, 162, 163, 169, 174
   car, 161, 171, 172, 177
   roll-on/roll-off, 74
   train, 171–2
Fertilisers, 55, 73, 129, 133, 161, 169, 191
Fiji, 150, 191, 196, 197
Finland, 60, 64, 169
f.i.o. (free in and out), 16, 141
Fjords, 162, 168
Flag of convenience, 65, 69, 75, 107, 122, 180
Flanders, 34, 35, 36, 38, 40
Flatirons, 163
Fluytschip, 45, 47
f.o.b. (free on board), 16, 97, 181, 183, 191
France, 33, 106, 107, 119, 144, 155, 163, 170
Freedom of the seas, 64–6

Freight rates, 34, 73, 79, 83, 84, 101, 102, 103, 104, 134, 144, 145, 155, 191
  bulk carriers, 141, 143
  coastal shipping, 161, 165
  determining, 85–6
  developing countries, 180–4, 187, 190, 196, 199
  fluctuations, 83, 88, 94, 95, 96–7
  liners, 101
  short sea trade, 176
  tankers, 117, 124, 125, 137–8

Galleons, 43
Galleys, 37, 38
Germany, 28, 35, 55, 56, 64, 65, 69, 106, 125, 137, 170, 171, 182, 185, 188
  coastal shipping, 163, 164, 165
Ghana, 186
Grain, 34, 36, 45, 50, 52, 54, 73, 91, 93, 94, 95, 98, 99, 103, 127, 132, 137, 138, 140, 141, 161, 163
  carriers, 140, 141
  trade, 25, 37, 92, 96, 98, 140, 141, 169
Great Lakes, 60, 94, 137, 140, 157
Greece, 25–7, 69, 75, 106, 107
Greenland, 34, 35, 61
Gulf of Aden, 26, 27
Gulf of Aqaba, 26, 66
Gulf of Bothnia, 28, 60, 95

Hafskip, 34
Haj, 187
Hanseatic League, 35, 45
Harwich, 172, 173, 174, 176
Holland, *see* The Netherlands
Home trade, 159, 161, 166, 171
Hovercraft, 159, 174
Hydrofoil, 159, 162

Ice, 59, 61, 94, 95, 147
Icebergs, 60
Iceland, 33, 34, 35, 45, 144
India, 23, 26, 27, 28, 38, 39, 42, 46, 50, 51, 70, 106, 112, 183, 187
  grain harvest, 94, 141
  national fleet, 186–7
  UK–India conference, 102, 187
Indian Ocean, 26, 27, 39, 40, 42, 43, 45, 49, 51, 58
Indonesia, 39, 41, 189, 198–9, 200
Insurance, 28, 49, 50, 56, 59, 60, 87, 117, 118, 119, 120, 121, 147, 152
Ireland, 28, 29, 33, 34, 35, 54, 105, 120, 159, 160, 161, 164
Iron, 34, 52, 131, 133, 147
  industry, 131, 134
  ore, 64, 71, 73, 132, 133–4, 136, 138, 140, 144
  ore carriers, 86, 134, 136, 145
  ore trade, 93, 97, 133, 137

Iron Age, 30
Israel, 62, 66, 124, 185, 187–8, 189
Italy, 36, 37, 38, 69, 106, 107, 112, 172, 185

Japan, 40, 55, 63, 66, 71, 108, 110, 150, 153, 154, 155, 185, 198
  shipbuilding, 77–8, 144–5
  steel industry, 141, 144, 145, 146
Java, 46, 198, 199
Junks, 40
Jute, 50, 99, 179, 191

Kiel Canal, 64, 65, 95, 163, 171

LAFTA, 81, 188, 189
Land bridge, 63, 157, 177
Lash ship, 16, 151, 153, 154, 200
Lateen rig, 16, 26, 36, 37, 39, 41
Latin America, 76, 81, 179, 188
Laying-up, 84–6, 97
Lebanon, 24, 25
Levant, 24, 25, 26, 33, 36, 37, 125
Liberia, 70, 75, 115
Lift-on/lift-off vessel, 150
Lloyd's, 50
LNG (liquefied natural gas), 128, 164
Loadline zones, 16, 67
Location of industry, 77, 113–14, 120, 131, 133, 134–6, 143, 146, 147, 157, 160
London, 45, 46, 47, 50, 51, 55, 56, 60, 62, 91, 141, 169, 170
  commercial growth, 48–9
Longships, 34
LPG (liquefied petroleum gas), 128, 164

Malaysia, 198, 199
SS *Manhattan*, 60
Maritime technology, 19–20, 27, 29–30, 36, 42, 52, 54, 56, 57, 61, 86, 105, 131, 132, 164, 167
  automation, 86, 87
  efficiency, 55, 69, 75, 116–17
  radar, 61
  telegraph, 55, 91
Masters, 34, 46, 59, 65, 193
Maury, Lt. M. F., 51
Meat trades, 55, 98, 105
Mediterranean Sea, 24, 25, 26, 27, 28, 29, 30, 32, 33, 34, 35, 42, 43, 62, 108, 116, 146, 154, 156, 159, 160, 166, 172
  pipelines, 123–4, 125
  shipping, 36–8
Mexico, 43, 188
Middle East, 33, 56, 62, 76, 110, 113, 127, 179, 188
Monopolies, 46, 49, 101, 102, 156
Monsoons, 26, 27, 39
  north-east, 24, 39
  north-west, 41

Index

south-east, 39, 41
south-west, 26, 39, 51, 58
Montreal, 60, 157

Natural gas, 128, 160, 168
Navigation, 39, 42, 43, 46, 51, 52, 61
Neolithic Age, 25, 28, 29
The Netherlands, 32, 45-6, 47, 49, 64, 106, 116, 147, 155, 165, 170, 185, 198
   vessels, 45, 163, 171
New South Wales, 51, 144
New York Exchange, 81
New Zealand, 20, 41, 55, 104, 105, 113, 119, 120, 144, 147
Nigerian National Shipping Line, 186
North America, 20, 34, 50, 52, 63, 91, 94, 95, 105, 110, 140, 150, 154, 155, 157, 159, 186
North Sea, 45, 49, 61, 64, 125, 147, 166, 167, 168, 169, 171, 174, 177
   natural gas, 128
Northern Way, 33
North-west passage, 60, 125
Norway, 28, 29, 33, 35, 56, 79, 116, 138, 159, 162-3, 166, 168, 169

OBO ships, 137, 138
OECD, 185
Oil, 56, 60, 61, 63, 65, 67, 91, 110, 112, 113, 114, 116, 121, 122, 127, 128, 133, 134, 136, 137, 138, 141, 160, 164, 169, 188, 199, 200
   crude, 73, 112, 113, 114, 116, 117, 120, 121, 124, 125, 127, 128, 160, 188
   industry, 82, 143, 165
   'Oil in Navigable Waters' Act, 65, 121
   trade, 62, 110, 125-8, 160
Oil refining, 112, 113, 121, 128
   refineries, 112-14, 119, 120, 121, 125, 171, 188
Oil tankers, 56, 62, 63, 74, 82, 92, 110-28, 131, 141, 145, 153, 165, 188, 189, 199
   crude, 113, 114, 115, 120, 127
   pipelines, 122-3, 124, 125
   pollution, 121-2
   product, 113, 114
   size, 114, 115, 116, 117
Ore, 72, 131, 133, 136, 137, 140, 145, 146, 147
Overseas Containers Ltd. (OCL), 148, 155
Overtonnaging, 69, 86, 117, 155, 175, 176, 185

Pacific Ocean, 40, 41-2, 43, 58, 59, 60, 63, 98, 108, 155, 179
   developing countries, 190, 195-8
Pakistan, 94, 183, 187, 191
Palletisation, 148, 153

carrier, 150-1, 153, 162, 164, 174
pallets, 148, 153, 200
Panama, 64, 137
   Canal, 63, 64, 65, 141, 148
   Isthmus, 202
PANHOLIB, 75, 115
Panimax ships, 63
Paragraph ships, 164, 171
Passenger ships, 34, 36, 56, 73-4, 99, 105-8, 161, 162, 163, 171, 188
   ferries, 172
   hovercraft, 174
   migration, 105, 107, 188
Persian Gulf, 23, 37, 116, 117, 120, 121, 123, 124, 128, 201
Petrochemicals, 113, 164
Philippines, 41, 189, 198, 201
Pipelines, 61, 120, 122-5, 133, 136, 137, 147, 160, 161
Poland, 64, 163, 168, 171, 185, 187
Pollution, 65, 66, 118-19, 121-2, 127, 128
Polynesia, 195
Portugal, 38, 42, 45, 46, 49, 106, 129
Pre-slinging, 148, 200
Proas, 40, 41
Productivity of shipping, 51, 153, 175

Railways, 48, 54, 55, 91, 95, 156, 157, 160, 163, 165, 171, 172, 174, 175, 176, 187
   British Rail, 165, 172, 174, 175, 176
   container ships, 173-4
Rebates, 102, 104, 186, 190
Red Sea, 24, 26, 27, 37, 39, 40, 65, 123, 188, 201
Refrigeration, 55, 98, 100, 128, 148, 161
Rhine, 32, 125, 171
Rochdale Report, 76, 156, 186, 187
Romans, 27, 30, 32, 33, 36, 38, 43
Roll-on/roll-off vessels, 150, 154, 164, 165, 170, 171, 172, 173, 174-5, 176, 200
Rotterdam, 140, 141, 161, 169, 170, 171, 173, 176
Russia, *see* U.S.S.R.

St. Lawrence River, 94, 140
St. Lawrence Seaway, 60, 65, 157
Scandinavia, 28, 35, 36, 69, 91, 133, 138, 145, 150, 168, 170, 171, 174, 177
Scotland, 28, 29, 34, 160, 161
Seamen, 24, 25, 26, 27, 30, 33, 37, 38, 39, 40, 42-3, 46, 48, 195
   and unitisation, 156-7
Seasons, 24, 26, 34, 39, 41, 47, 58-9, 67, 93-5, 100, 106, 125-7, 137
Shipbuilding, 23-7, 29, 30, 32-3, 34, 36, 37-8, 42, 43, 45
   British, 47, 49, 146
   Japanese, 77-8

tankers, 114, 116, 117, 120
Shipbuilding loans, 76, 185
Shipowners, 27, 34, 35, 59, 62, 65, 76, 77, 78, 81, 82, 86, 88, 138, 151, 169, 171
  organisation, 81, 83
  tankers, 114, 115–16, 117, 118, 120, 122
  tramps, 90, 92, 96, 97, 98
  unitised shipping, 151, 156
Shipping costs, 41, 59, 78, 82, 84–8, 92, 97, 145, 151
  bulk carriers, 131, 134, 136, 137, 140, 143
  coastal shipping, 161, 162
  crew costs, 75, 86, 87, 91, 92, 152, 162, 192
  developing countries, 185–6, 188, 189, 192
  liners, 100–1, 103–4, 105
  pipelines, 122–3, 124, 125
  tankers, 117–18, 120, 122, 128
  transport, 140, 143, 145, 147, 157, 166, 169, 176
  unitisation, 151, 152, 153, 157
Shipyards, 47, 77, 117, 187
Short sea trade, 150, 153, 159, 164, 166–77
  cargoes, 168–70
  developing countries, 191, 192, 201
  passengers, 171
  systems, 174–6
Side ports, 174
Singapore, 113, 199, 200, 201
  Strait, 61
Slave trade, 40, 50, 65
Slurry, 133, 137, 147
South America, 20, 43, 50, 71, 91, 107, 133, 145, 146, 159, 188–9
Spain, 33, 34, 37, 42–3, 45, 49, 133, 172
Specialised ships, 74–5, 92, 128–9, 137, 140, 169–70, 171, 183, 202
Standing off-and-on, 16, 192, 195
Steamships, 52–4, 55, 62, 90, 91
Steel, 133, 136, 140, 143, 145, 146
  industry, 83, 131, 133, 134, 141, 144, 146, 147
Stowage factor, 16, 136, 191
Strait of Dover, 61, 73, 176, 202
Strait of Gibraltar, 26, 65
Strait of Malacca, 39, 61, 199
Strait of Tiran, 66
Subsidies, 56, 69, 75, 76, 84, 105, 107, 155, 161, 185, 186, 190
Suez, 113, 188
  Canal, 54, 60, 62, 63, 65, 71, 90, 107, 114, 117, 120–1, 124, 125
  Gulf, 24, 25
  Isthmus, 42, 62, 125, 202
Sugar, 50, 73, 83, 133, 137, 199

Sulphur, 127, 129, 133, 170
Sweden, 28, 29, 33, 60, 64, 66, 155, 163, 169, 177, 185

Tankers, 86, 140, 141, 154, 160
  special, 128–9, 170
Tides, 61, 120
Timber, 24, 36, 39, 45, 47, 51, 54, 72, 73, 95, 99, 133, 143, 163, 169, 191, 199, 200, 201
  carriers, 103, 137, 143, 189
  trade, 25, 45, 52, 93, 95, 143, 163
Tonnage, 16, 44, 47, 51, 54, 55, 69, 70, 76, 78, 82, 83, 85, 86, 87, 88, 91, 97, 133–4
  bulk carriers, 132, 133, 134, 136, 140, 141, 145
  cargo, 71–2, 73, 75, 94, 96, 100
Trepang, 17, 41
Tumble home, 17, 45
Turnround, 74–5, 88, 110, 120, 136, 150, 151–2, 153, 169, 200
Tyne, 45, 47, 48, 91, 146

UNCTAD, 153, 189
Unit loads, 137, 152, 171, 200
United States, 55, 56, 61, 63, 64, 65, 69, 70, 75, 76, 96, 104, 107, 108, 129, 137, 141, 154, 155, 156, 164, 169, 187
  east coast, 125, 140, 146, 156
Unitisation, 147–8, 151, 154, 156, 157
  effect on labour, 156–7
Unitised shipping, 131, 153, 156, 169, 171, 174
  vessels, 73, 74, 148, 150, 152, 154, 164, 171, 200
U.S.S.R., 33, 54, 64, 69, 76, 94, 104, 106, 107, 116, 157, 187, 198

Venice, 35, 36, 37, 38
Vikings, 33–6, 39
VLCC, 117, 118, 119, 120, 122, 124, 125

WALCON, 186
Wales, 28
  South Wales, 48, 134, 146
Waves, 58–9
Weather, 30, 34, 35, 38, 39, 40, 47, 61, 84, 94, 95, 120
  routeing, 59, 67
West Indies, 43, 49, 50, 51, 59
White Sea, 33, 34, 95
Wine, 45, 47, 129
Wool, 34, 35, 39, 45, 46, 50, 52, 157, 179

Yugoslavia, 185, 187

Zim Line, 186, 189

For Product Safety Concerns and Information please contact our
EU representative GPSR@taylorandfrancis.com Taylor & Francis
Verlag GmbH, Kaufingerstraße 24, 80331 München, Germany